Electricity 2

DEVICES
CIRCUITS
MATERIALS

THIRD EDITION

Electricity 2

DEVICES CIRCUITS MATERIALS

THIRD EDITION

THOMAS S. KUBALA

LIBRARY OF CONGRESS CATALOG CARD NUMBER: 79-93323
ISBN: 0-8273-1359-4

10 9 8 7 6 5 4 3 2

Printed in the United States of America
Published simultaneously in Canada by
Delmar Publishers, a division of
Van Nostrand Reinhold, Ltd.

DELMAR PUBLISHERS INC. • ALBANY, NEW YORK 12205

Preface

Electrical energy is the most important and widely used form of energy today. The electrician provides the specialized expertise and techniques required to put electricity to work for the benefit of all of us. The electrician must have a thorough knowledge of electrical devices and materials, and must understand the theory and characteristics of electrical circuits. The electrician is expected to apply this knowledge to practical circuit installations such as those found in homes, businesses, and industries.

ELECTRICITY 2 enables the student to learn the many characteristics of various types of alternating-current circuits and the devices that make up the circuits. In addition, this text serves as the basis for further study. The student should realize that the study of electricity is a continuing process. The electrical industry is constantly introducing new and improved devices and materials, while electrical codes undergo periodic revisions to upgrade safety and quality in electrical installation. While studying the units in this text, the student will find it valuable to refer to the most recent edition of the National Electrical Code (NEC), published by the National Fire Protection Association in the interest of life and property protection. State and local regulations, as they apply to a specific area, should also be consulted when making an actual installation.

The emphasis in content of ELECTRICITY 2 is on the practical applications of electrical concepts and the use of materials in functional electrical systems. The student is advised that electron movement (from negative to positive) is used in this text to define current direction.

The text is easy to read and comprehend, and the topics are presented in a logical sequence. The mathematical level involves algebra and the basic elements of trigonometry. No prior knowledge of trigonometry is assumed, and the student should not find it difficult to understand and calculate the simple mathematical equations required in this study.

Units 1–3 present the basic principles of alternating current. These units assist the student in developing an understanding of fundamental circuit devices. Units 4–8 explain various circuit configurations using different combinations of devices. Unit 9 presents the concept of power, units 11–13 cover practical situations that an electrician often encounters, and units 14 and 15 deal with fluorescent lighting. Summary Review units 10 and 16 are comprehensive examinations of previous units. The purpose of these review units is to evaluate the knowledge and understanding of the theory acquired by the student to that point. The Achievement Review at the end of each of the other units requires thought and comprehension on the part of the student. These reviews are essential to the learning process required by this text.

All students of electricity will find this text useful, especially those in electrical apprenticeship programs, trade and technical schools, and various occupational programs.

An Instructor's Guide is available. An outline of each unit is provided, including suggested class demonstrations, to assist the instructor in planning the presentation to the students. In addition, answers and solutions are provided for each Achievement Review and the two Summary Reviews. Complete solutions for the more complex problems are also included.

Dr. Thomas S. Kubala, the author of ELECTRICITY 2, is the President of Thomas Nelson Community College, Hampton, Virginia. Prior to accepting this post, he was Dean of the College, Anne Arundel Community College, Arnold, Maryland, having previously held such positions there as Dean of Career Programs, Associate Professor of Engineering, and Director of Electrical Technology.

Dr. Kubala received an AAS degree in Electrical Technology from Broome Technical Community College, Binghamton, New York. He earned a BS degree in Electrical Engineering from the Rochester Institute of Technology, Rochester, New York, and an MS degree in Vocational-Technical Education from the State University of New York at Oswego, Oswego, New York. He earned his doctoral degree from the University of Maryland, College Park, Maryland.

In addition to his extensive background in technological education, Dr. Kubala has also had several years of experience in industry with responsibilities in the fields of aerodynamics, and electrical drafting, design, testing, and evaluation. He has been a member of several professional organizations during his career.

Dr. Kubala is also the author of ELECTRICITY 1, PRACTICAL PROBLEMS IN MATHEMATICS FOR ELECTRICIANS, and CIRCUIT CONCEPTS: DIRECT AND ALTERNATING CURRENT, all published by Delmar.

A current catalog including prices of all Delmar educational publications is available upon request. Please write to:

Catalog Department
Delmar Publishers Inc.
50 Wolf Road
Albany, New York 12205

ELECTRICAL TRADES

The Delmar series of instructional material for the electrical trades consists of the texts, text-workbooks, laboratory manuals, and related information workbooks listed below. Each text features basic theory with practical applications and student involvement in hands-on activities.

- ELECTRICITY 1
- ELECTRICITY 2
- ELECTRICITY 3
- ELECTRICITY 4
- ELECTRIC MOTOR CONTROL
- ELECTRIC MOTOR CONTROL LABORATORY MANUAL
- ALTERNATING CURRENT FUNDAMENTALS
- DIRECT CURRENT FUNDAMENTALS
- ELECTRICAL WIRING – RESIDENTIAL
- ELECTRICAL WIRING – COMMERCIAL
- ELECTRICAL WIRING – INDUSTRIAL
- PRACTICAL PROBLEMS IN MATHEMATICS FOR ELECTRICIANS

EQUATIONS BASED ON OHM'S LAW

P = Power in Watts

I = Intensity of Current in Amperes

R = Resistance in Ohms

E = Electromotive Force in Volts

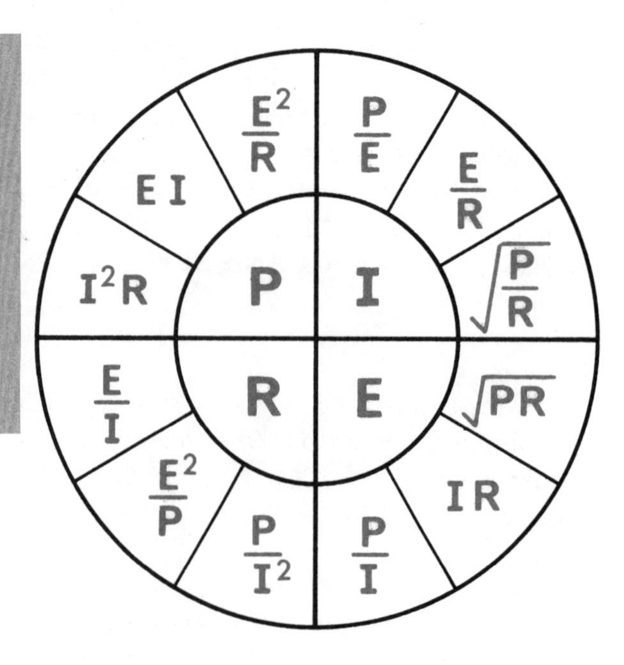

Contents

The author and editorial staff at Delmar Publishers are interested in continually improving the quality of this instructional material. The reader is invited to submit constructive criticism and questions. Responses will be reviewed jointly by the author and sponsoring editor. Send comments to:

DELMAR PUBLISHERS INC.
Sponsoring Editor
Technical Division
50 Wolf Road
Albany, NY 12205

1 Alternating-Current Principles

OBJECTIVES

After studying this unit, the student will be able to

- discuss the characteristics of alternating current.
- describe the generation of alternating current.
- define the terminology related to alternating current.

Most of the electrical energy used in the United States is generated as alternating current (ac). This is not because alternating current is superior to direct current (dc) in its industrial and residential applications. In fact, there are many instances where direct-current energy is necessary for industrial purposes. Where direct current is required, alternating current is generated in a power station, transmitted some distance, and then converted to direct current at the point where it is to be used.

The reasons for generating nearly all electrical energy as alternating current are as follows:

1. Alternators (ac generators) have no commutators. Therefore, units having larger power ratings and the resultant heavier current ratings can be used without the problem of brush arcing and heating.
2. Since alternators do not have commutators, they are capable of generating comparatively high voltages, such as 11 000 to 13 800 volts.

Hydroelectric Generators

35 555-kVA, 32 000-kW, 13 800-V, 105.8-r/min generators

Turbine-driven Generator

15 000-kW, 3 600-r/min steam turbine with hydrogen-cooled generator

Fig. 1-1 Types of generators. (*Courtesy Allis-Chalmers Manufacturing Company*)

3. Alternating-current energy may be transmitted economically over great distances. Therefore, alternating current can be generated in large quantities in a single station and distributed over a large territory.
4. For constant-speed work, the alternating-current, squirrel-cage induction motor is less expensive than the direct-current motor, both in initial cost and in maintenance.

GENERATION OF ALTERNATING VOLTAGE

The volt (V) is the unit of electromotive force. One volt is developed by cutting 100 million magnetic lines of force in one second.

The simplest method of generating electromotive force is by turning a single loop of wire between two magnetic field poles.

Figure 1-2 illustrates a simple alternating current generator. The left hand rule is used to determine the direction of the current in the coil and in the external circuit created by the generated electromotive force.

Figure 1-3 is a more convenient form of representing the simple generator in figure 1-2. A front-view section from the slip ring side of the generator is shown.

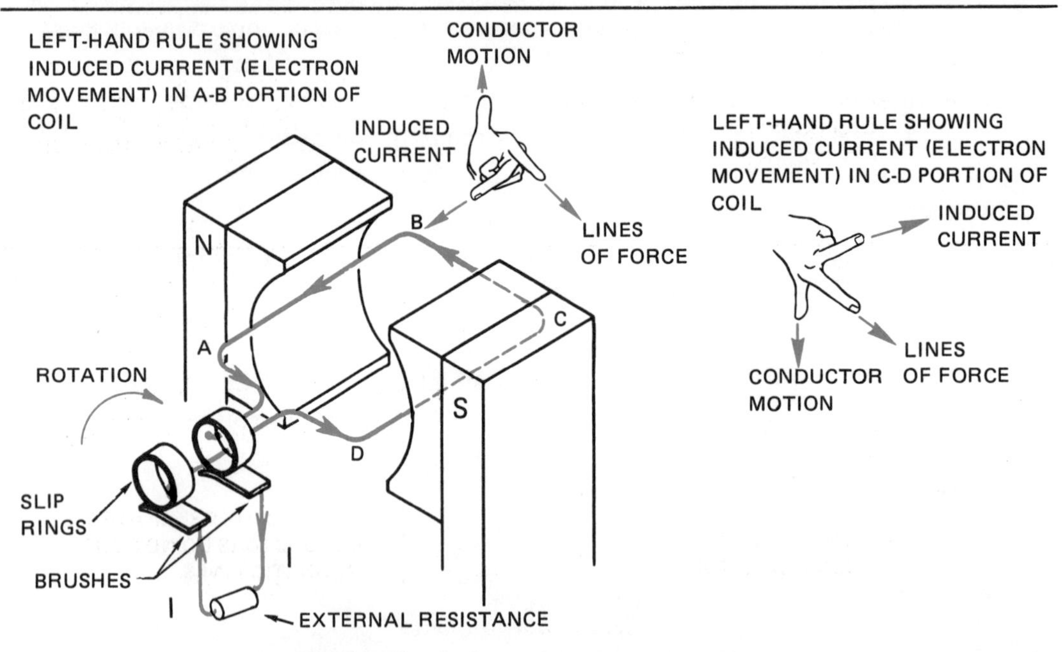

Fig. 1-2 Simple alternating-current generator.

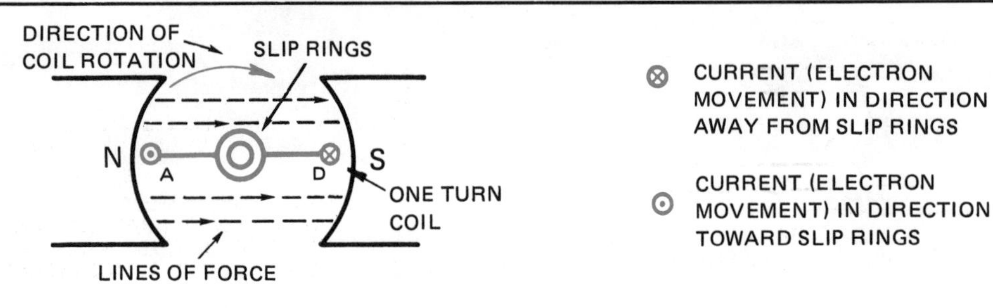

Fig. 1-3 Simple generator.

DEVELOPMENT OF A CYCLE

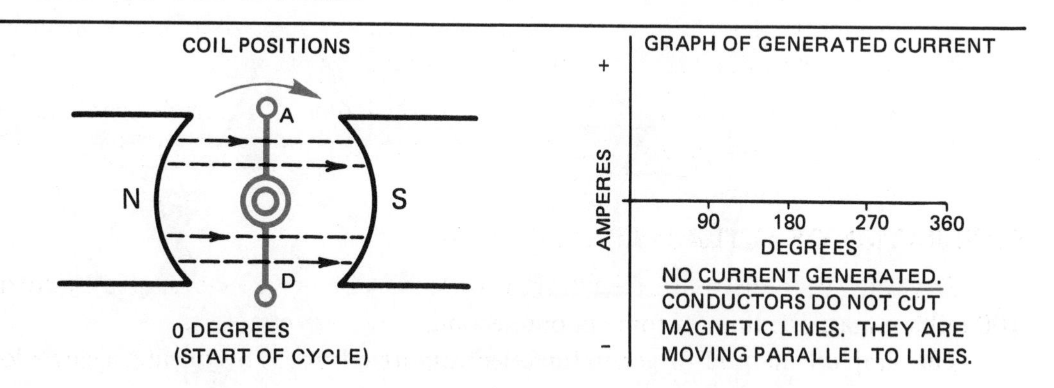

Fig. 1-4 Start of cycle.

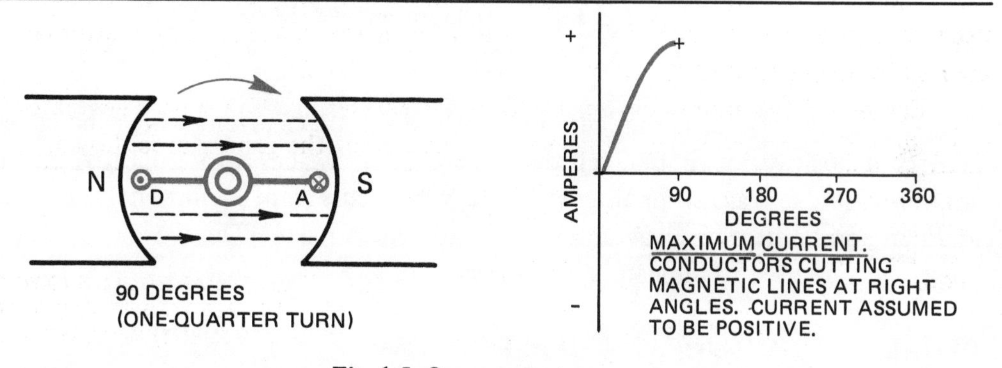

Fig. 1-5 One-quarter turn.

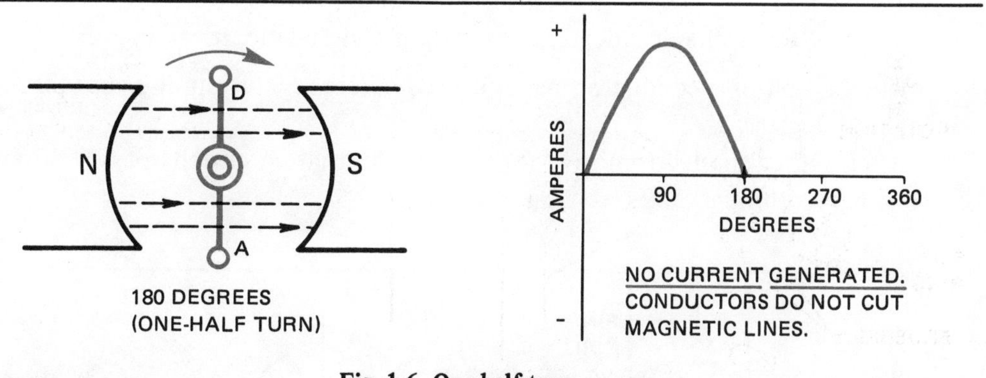

Fig. 1-6 One-half turn.

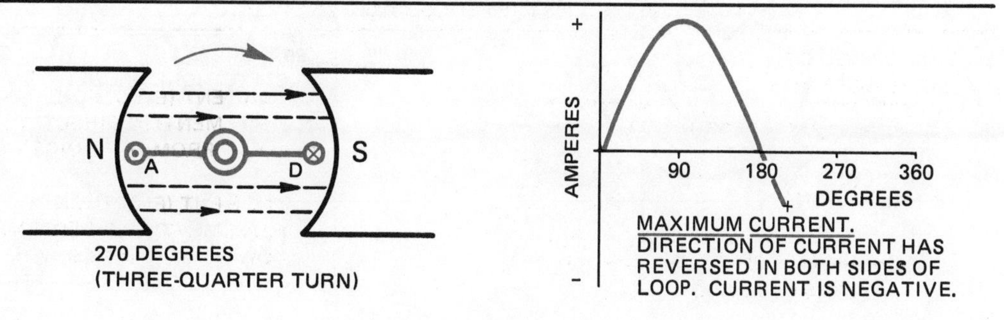

Fig. 1-7 Three-quarter turn.

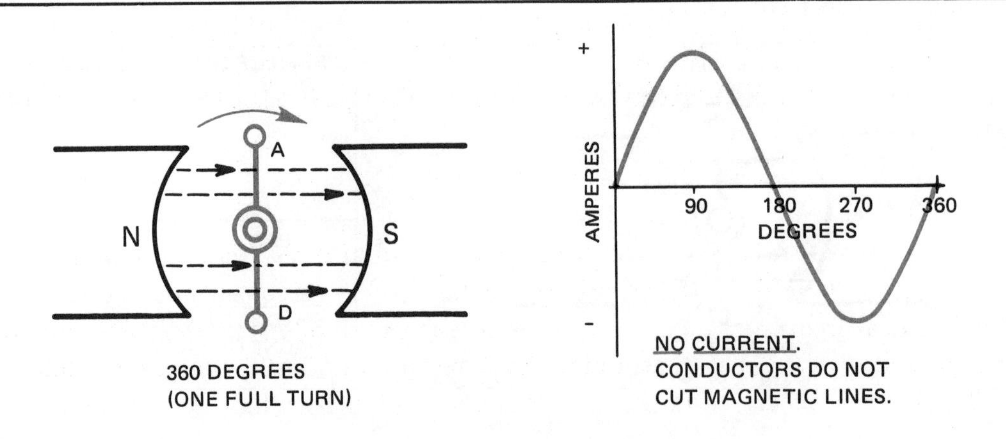

Fig. 1-8 One full turn completed.

Cycle

One *cycle* of alternating current is defined as that current which increases from zero to a positive maximum, returns to zero, then increases to a negative maximum, and returns to zero again. In other words, a cycle occurs from a point on the waveform to another point where the waveform begins to repeat itself. The words "positive" and "negative" are used to indicate opposite directions.

DEGREES – MECHANICAL AND ELECTRICAL

In figure 1-8, the complete turn represents:

360 mechanical degrees of rotation and 360 electrical degrees

When a coil or a conductor makes one complete revolution, it passes through 360 mechanical degrees.

When either an electromotive force or an alternating current passes through one cycle, it passes through 360 electrical degrees.

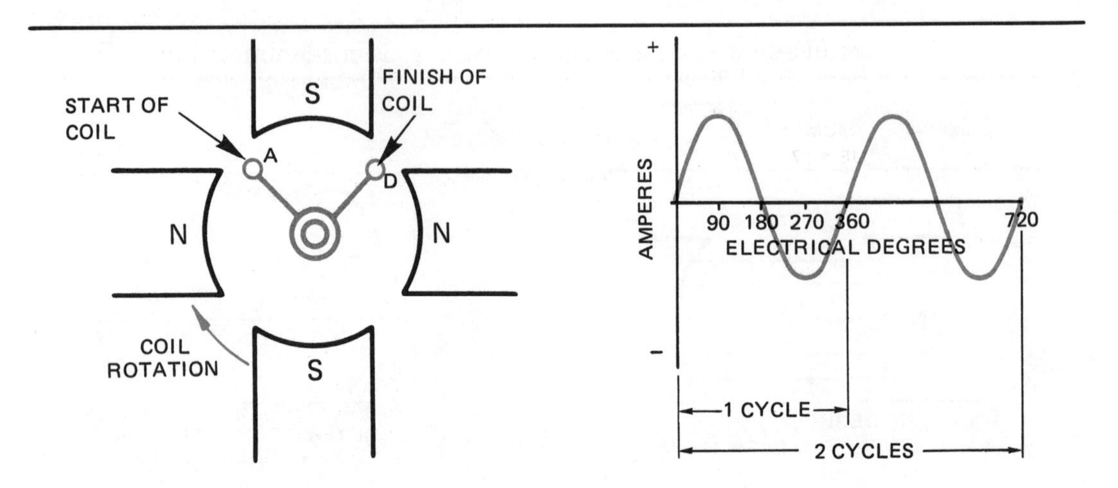

Fig. 1-9 A four-pole generator – two cycles per mechanical revolution.

If a single-loop coil is placed in a generator having four magnetic poles, two complete cycles of alternating current or 720 electrical degrees are generated in one mechanical revolution, since one cycle is generated when each side of a coil or loop passes two poles.

FREQUENCY

If the coil in figure 1-8 turns at the rate of 3 600 revolutions per minute (3 600 r/min), or 60 revolutions per second, 60 electrical cycles are generated in one second. The electrical frequency then is 60 cycles per second, or 60 hertz (Hz). With the four-pole generator in figure 1-9, assuming the same speed of 3 600 r/min, the number of electrical cycles generated in one second is 120. This is true since two electrical cycles are generated during each mechanical revolution. The frequency is then 120 Hz.

Formulas commonly used to find the frequency, speed, and number of poles are:

$$F = \frac{NP}{120} \qquad N = \frac{120F}{P} \qquad P = \frac{120F}{N}$$

Where
- F = frequency in hertz
- N = speed in r/min
- P = number of poles

EFFECTIVE VALUE OF ALTERNATING CURRENT

A 60-Hz, 120-volt line has current and electromotive force varying from zero to maximum positive and negative values 60 times each second. Instantaneous values are of little value. The effective values, figure 1-10, have been accepted as practical working quantities. When values of current and voltage are specified in ac circuits, it is understood that they are effective values, unless otherwise specified. The values indicated by ac ammeters and voltmeters are effective values.

The *effective value* of alternating current is that value which produces the same amount of heat as the same value of direct current.

Example: An electric heater is rated at 10 amperes (A) ac or dc. Either current produces the same amount of heat in a given amount of time.

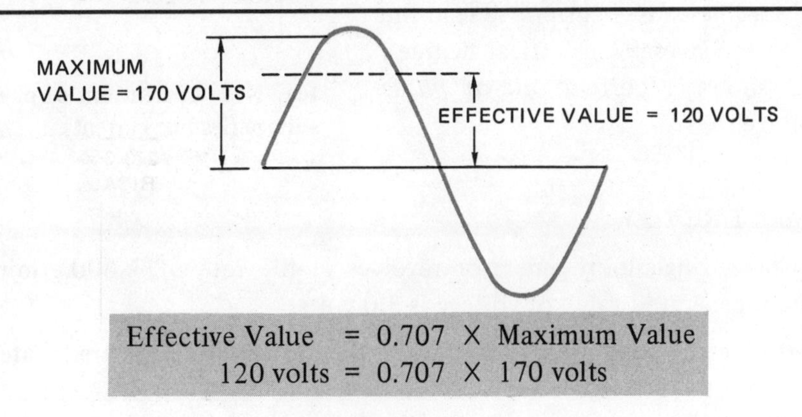

Effective Value = 0.707 X Maximum Value
120 volts = 0.707 X 170 volts

Fig. 1-10.

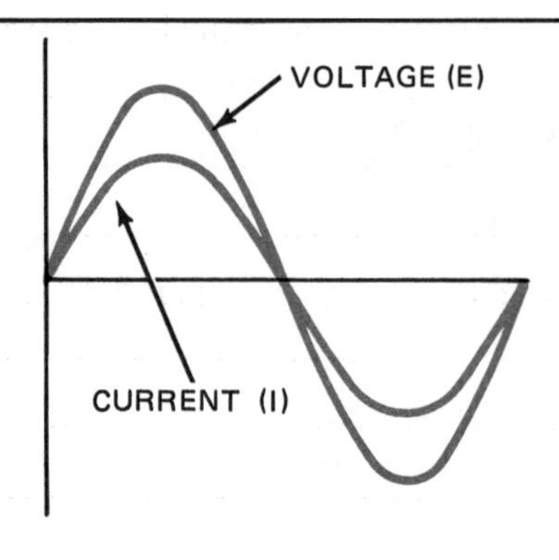

Fig. 1-11 Current in phase with the voltage.

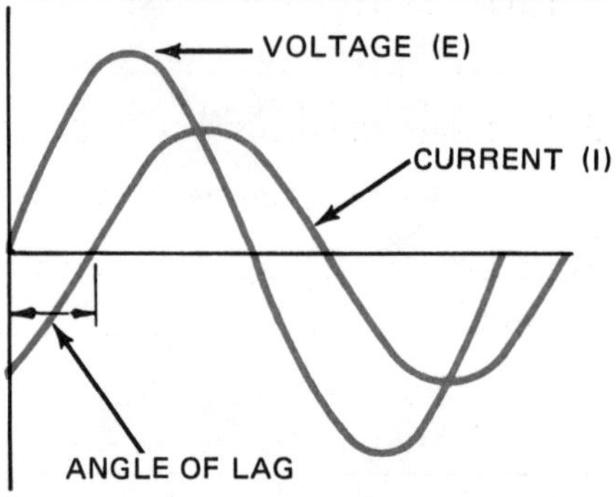

Fig. 1-12 Current out of phase with the voltage (lagging current).

SINGLE PHASE

A voltage that is generated in a single winding of a generator is called *single-phase voltage.*

IN PHASE

When both the voltage wave and the current wave reach their corresponding zeros, maxima, and intermediate values at exactly the same time, they are said to be *in phase,* as shown in figure 1-11.

OUT OF PHASE – LAGGING

When current is supplied to an induction motor, it lags the voltage. This current is *out of phase* with the voltage, as shown in figure 1-12.

OUT OF PHASE – LEADING

A synchronous motor or a capacitor connected to a line causes current to lead the voltage by as much as 90 electrical degrees. Figure 1-13 shows a current *out of phase* with the voltage.

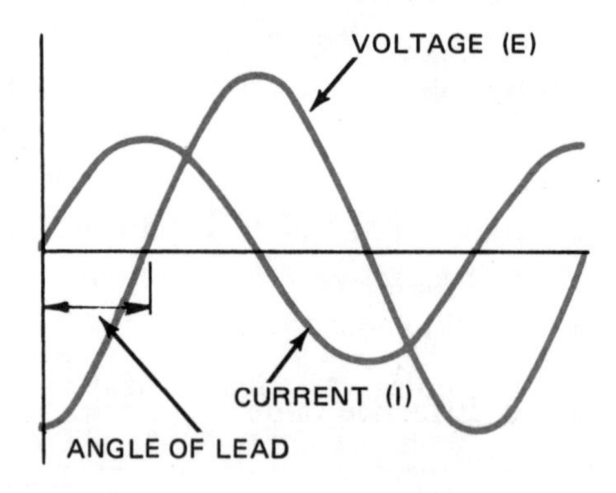

Fig. 1-13 Current out of phase with the voltage (leading current).

ACHIEVEMENT REVIEW

1. A four-pole, single-loop generator revolves at the rate of 3 600 r/min, and the maximum generated value of voltage is 3.0 volts.
   a. Determine the generated voltage when the loop conductors are located in front of the poles. (See figure 1-5) ____________________
   b. Determine the generated voltage when the loop conductors are located between the poles. (See figure 1-6) ____________________

c. How many electrical cycles are generated in one mechanical revolution? ______________________

d. Calculate the electrical frequency. ______________________

e. Calculate the effective value of voltage. ______________________

2. State four advantages of generating alternating current as compared to direct current. ______________________

____

____

____

3. An ammeter indicates 15 amperes in an ac induction motor line.
   a. What value is measured? (effective) (instantaneous) (average)

   ____

   b. Underline the correct phase in the following sentence: The phase of this current is (in phase) (out of phase – lag) (out of phase – lead).

4. How many electrical and mechanical degrees are there in one complete mechanical revolution for each of the specified generators?

Generator	Mechanical	Electrical
2 pole	________	________
4 pole	________	________
6 pole	________	________

5. Define each of the following terms:
   a. Cycle ______________________

   ____

   b. Frequency ______________________

   ____

   c. Effective value ______________________

   ____

   d. In phase ______________________

   ____

6. A 10-pole alternator revolves at 600 r/min. What is the value of the electrical frequency generated? ______________________

7. To produce a frequency of 25 Hz, at what r/min must a 6-pole alternator be driven? ______________________

8. To produce a frequency of 60 Hz, how many poles does an alternator have if the speed is 150 r/min? ______________________

2 Inductance and Inductive Reactance

OBJECTIVES

After studying this unit, the student will be able to

- describe an inductive circuit.
- describe what is meant by self-induction and mutual induction.
- define inductive reactance.
- demonstrate the relationship between voltage and current in various inductive circuits by the use of vectors.

A coil of wire is an important part of many pieces of electrical equipment. A magnetic field is produced when current exists in the coil. As the strength of the magnetic field changes, an induced electromotive force (emf) is created across the coil. The induced voltage opposes the source voltage. As the opposition becomes greater, less current exists in the circuit.

The coil has a property that opposes any change in current. This property is called *inductance* (L). The amount of opposition to current change is called *inductive reactance*, and is a function of frequency and inductance.

LENZ'S LAW

According to *Lenz's law*, the induced voltage in a coil is always in such a direction as to oppose the effect that produces it.

Self-Inductance

When the varying lines of magnetic force induce an electromotive force in the coil itself, the coil has *self-inductance.*

Mutual Inductance

When the varying lines of magnetic force from a coil induce an electromotive force in an adjacent coil, the coils have *mutual inductance.* Figure 2-1 illustrates a transformer containing a primary coil and a secondary coil. The primary coil has a current in it that creates a magnetic

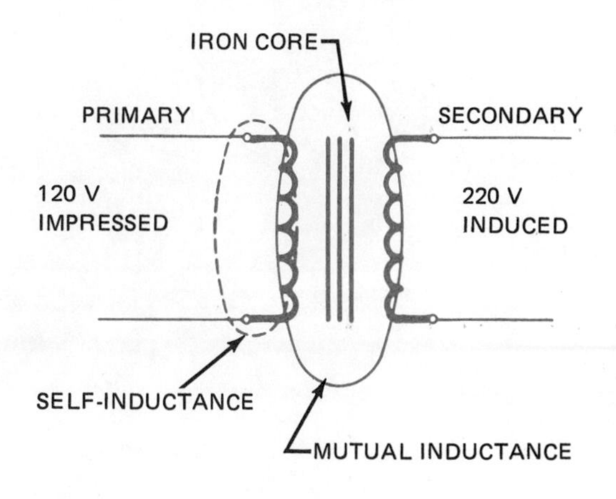

Fig. 2-1 Transformer showing location of self-inductance and mutual inductance.

field. Part of the field links the secondary coil. Since the field is *changing*, a voltage is induced in the secondary. This is a *step-up transformer* because the secondary voltage is greater than the primary voltage.

Measurement of Inductance

The unit of inductance is the henry (H). A circuit or coil has an inductance of one henry when current varying at the rate of one ampere per second induces an emf of one volt across the terminals of the circuit or coil. The inductance can be varied by varying the amount of magnetic flux or the number of *turns* in the coil.

Effect of Inductance

Connect a lamp in series with a coil having a movable iron core. Connect the combination to an ac supply. Note the following conditions:

Core *out* of coil	-	Lamp will be bright
Core *in* coil	-	Lamp will be dim

When the core is out of the coil, few lines of magnetic force are produced by the coil since air is a poor magnetic conductor. The induced electromotive force is weak, and little opposition is offered to the line voltage. Therefore, a normal quantity of current exists in the lamp.

When the iron core is inserted in the coil, a better magnetic path is provided, there is a higher induced electromotive force and, consequently, there is less current, as indicated by the dim lamp.

1.5 kVA, 93 V ac, 3 PHASE, 60 Hz, WEIGHT 60 POUNDS

Fig 2-2 A transformer. (*Courtesy General Electric Company*)

Inductive Reactance

The opposition in coils having inductance may be measured in ohms (Ω). If the frequency and inductance are known, the opposition, or *inductive reactance* (X_L) may be calculated.

$X_L = 2\pi fL$ Where: X_L = inductive reactance in ohms
π = 3.14
f = frequency in hertz
L = inductance in henrys

Examples:

Coil A f = 60 Hz L = 0.1 H
$X_L = 2 \times 3.14 \times 60 \times 0.1 = 37.7\ \Omega$

Coil B f = 60 Hz L = 0.2 H
$X_L = 2 \times 3.14 \times 60 \times 0.2 = 75.4\ \Omega$

If Coil B is connected to 120 Hz:
$X_L = 2 \times 3.14 \times 120 \times 0.2 = 150.7\ \Omega$

Therefore, it can be said that:

INDUCTIVE REACTANCE VARIES DIRECTLY WITH INDUCTANCE AND FREQUENCY

Lags of Current Due to Inductance

Tests show that if a coil having negligible resistance is connected to an ac line, the current lags the voltage by 90 degrees, as shown in figure 2-3.

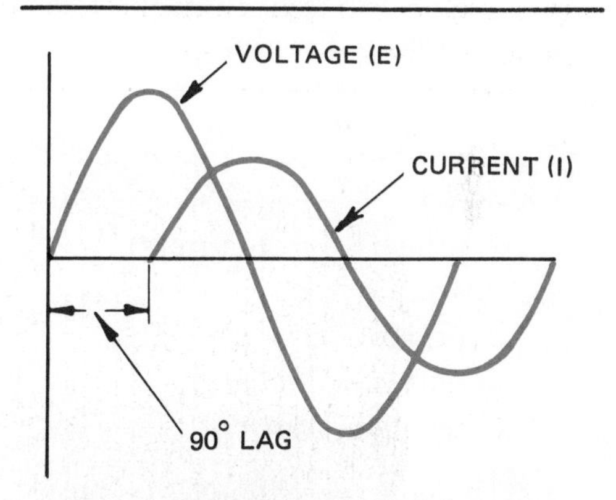

Fig. 2-3 Current lagging the voltage by 90 degrees.

VECTOR REPRESENTATION

The relationship between voltage and current in an inductive circuit is shown more conveniently by the use of vectors.

A vector is a line representing quantity or magnitude and direction. The vectors shown in figure 2-4 represent 110 volts with a current of 10 amperes lagging by 90 degrees. These vectors may be visualized as clock hands rotating in a counterclockwise direction.

The lengths of the vectors in figure 2-4 depend on the scale used. The scale for the voltage vector is 1 inch = 50 volts. The length of the vector is 110/50 = 2.2 inches. The scale for the current vector is 1 inch = 10 amperes.

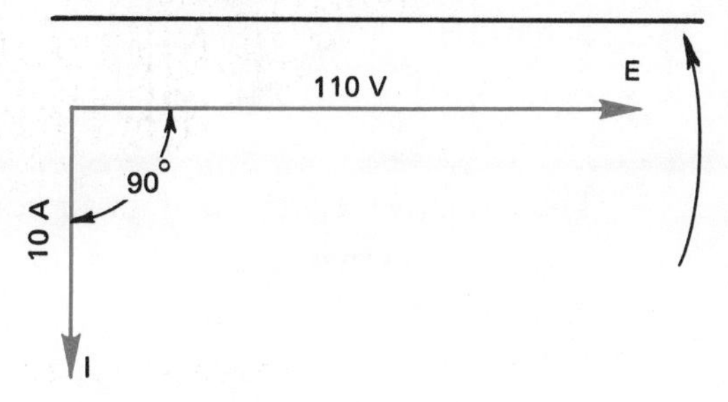

Fig. 2-4 Vector diagram of current lagging voltage by 90 degrees.

FINDING CURRENT

Figure 2-5 and the example show how current is determined in an ac inductive circuit.

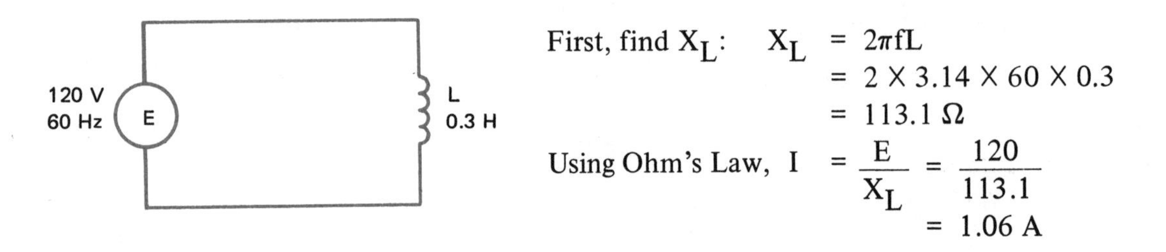

First, find X_L: $X_L = 2\pi fL$

$$= 2 \times 3.14 \times 60 \times 0.3$$

$$= 113.1\ \Omega$$

Using Ohm's Law, $I = \dfrac{E}{X_L} = \dfrac{120}{113.1}$

$$= 1.06\ \text{A}$$

Fig. 2-5 Inductive circuit.

ACHIEVEMENT REVIEW

In items 1–10, select the *best* answer to make the statement true. Place the letter of the selected answer in the space provided.

1. Inductance is __________
   a. the same as reactance.
   b. the property of a coil.
   c. magnetic field strength.
   d. measured in ohms.
   e. dependent upon reactance.

2. The amount of voltage induced in a transformer secondary coil is a function of __________
   a. primary direct current.
   b. a static field.
   c. primary self-inductance.
   d. mutual inductance.
   e. secondary current.

3. The unit for inductive reactance is the __________
   a. ohm.
   b. henry.
   c. hertz.
   d. vector.
   e. Lenz.

4. The inductive reactance of an air core coil may be increased by __________
   a. decreasing frequency.
   b. increasing source voltage.
   c. inserting an iron core.
   d. increasing current.
   e. decreasing inductance.

5. In a purely inductive circuit (no resistance), ____________
   a. current lags voltage.
   b. voltage lags current by 90°.
   c. current and voltage are in phase.
   d. voltage leads current.
   e. current lags voltage by 90°.

6. The inductive reactance of a 0.06-H coil connected to a 120-V, 60-Hz source is ____________
   a. 2.26 Ω.
   b. 3.6 Ω.
   c. 7.2 Ω.
   d. 22.62 Ω.
   e. 432 Ω.

7. A purely inductive circuit contains a voltage source of 280 V at 40 Hz. The total inductive reactance of the circuit is 20 Ω. The value of the total current, in amperes, is ____________
   a. 0.056.
   b. 0.08.
   c. 2.0.
   d. 7.0.
   e. 14.0.

8. A current of 5 A exists in a purely inductive circuit connected to a 120-V, 60-Hz source. The total inductive reactance of the circuit is ____________
   a. 0.064 H.
   b. 12.0 Ω.
   c. 24.0 Ω.
   d. 300 H.
   e. 600 Ω.

9. A 0.265-H coil is connected to a 250-V, 60-Hz source. The total circuit current, in amperes, is ____________
   a. 2.5.
   b. 4.17.
   c. 100.
   d. 226.
   e. 940.

10. A coil with a negligible resistance draws 7 A when connected to a 110-V, 25-Hz source. The inductance of the coil, in henrys, is ____________
    a. 0.064.
    b. 0.1.
    c. 3.57.
    d. 4.4.
    e. 15.7.

3 Capacitance and Capacitive Reactance

OBJECTIVES

After studying this unit, the student will be able to

- discuss the characteristics of capacitance.
- describe the effect of capacitance in an alternating-current circuit.
- use vectors to show the voltage and current relationship in a capacitor.

Practically all electrical equipment contains a combination of resistors or coils. Some industrial equipment, such as capacitor motors, capacitor banks and automatic switchgear, uses capacitors. Transmission lines have capacitance between the wires.

A capacitor consists of two plates of electrical conducting material separated by an insulating material. The plates are commonly aluminum, tin, or any other nonmagnetic substance. The insulating material, called the *dielectric*, may be any of a large variety of substances, such as air, mica, glass, oil, wax, paper, fiber or rubber.

When electric potential is connected to the plates, an electrical charge is stored in the capacitor. In an ac circuit, the alternating voltage causes the capacitor to charge and discharge during every cycle. Although current cannot pass through the capacitor, an ammeter connected in the line will measure current resulting from the alternating charge and discharge.

CAPACITANCE

Capacitance is the property of a capacitor, and is defined as the amount of electrical charge that a capacitor receives for each volt of applied potential. The unit for capacitance is the *farad* (*F*). However, the farad is a very large unit in terms of the charges that are normally present, so the *microfarad* (*μF*) is generally used.

$$1 \text{ farad} = 1\,000\,000 \text{ microfarads}$$

or

$$1 \text{ microfarad} = \frac{1}{1\,000\,000} \text{ farad}$$

CAPACITIVE REACTANCE

A capacitor in a circuit limits the current in a circuit just as resistance and inductive reactance limit current. The opposition from a capacitor is called *capacitive reactance* (X_C). If the capacity (in microfarads) and the frequency are known, the reactance in ohms may be calculated.

$$X_c = \frac{1\,000\,000}{2\pi fC}$$

Where:
X_c = capacitive reactance in ohms
π = 3.14
f = frequency in hertz
C = number of microfarads

If C is the number of farads, then the equation becomes:

$$X_c = \frac{1}{2\pi fC}$$

Examples:

Capacitor A
f = 60 Hz
C = 13 μF

$$X_c = \frac{1\,000\,000}{2 \times 3.14 \times 60 \times 13} = 204\ \Omega$$

Capacitor B
f = 60 Hz
C = 26 μF

$$X_c = \frac{1\,000\,000}{2 \times 3.14 \times 60 \times 26} = 102\ \Omega$$

Capacitor C
f = 120 Hz
C = 26 μF

$$X_c = \frac{1\,000\,000}{2 \times 3.14 \times 120 \times 26} = 51\ \Omega$$

Summary

CAPACITOR	f (Hz)	C (μF)	X_c (Ω)
A	60	13	204
B	60	26	102
C	120	26	51

CAPACITIVE REACTANCE VARIES INDIRECTLY WITH CAPACITANCE AND FREQUENCY

Capacitive reactance may be *decreased* by increasing either the capacitance or the frequency.

DANGER

A capacitor holds a charge for a long period of time following use in a circuit. DISCHARGE A CAPACITOR BEFORE HANDLING.

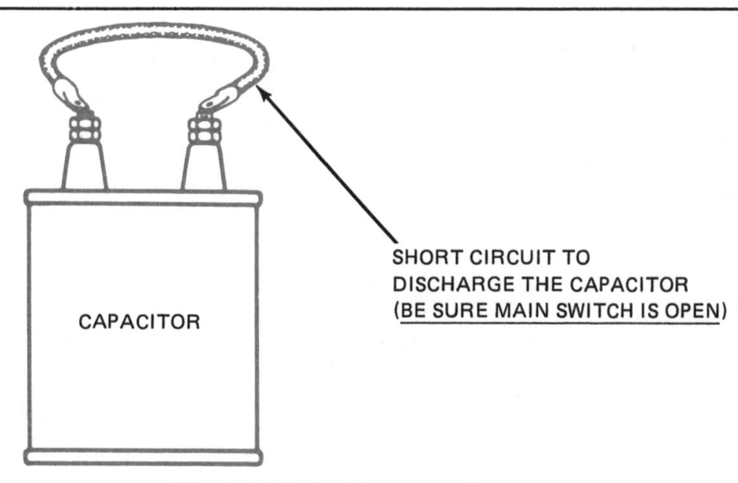

Fig. 3-1 Discharging a capacitor.

CURRENT LEADS THE VOLTAGE IN A CAPACITOR

A capacitor connected to an ac line causes the current to lead the voltage by 90 degrees, as shown in figure 3-2. Oscilloscope pictures of the current and voltage waveforms show this relationship. Figure 3-3 uses vectors to show the same information as figure 3-2.

When capacitors are connected in parallel, their combined capacitance may be found using the method by which the combined resistance of series-connected resistors is found:

$$C_t = C_1 + C_2 + C_3$$

When capacitors are connected in series, their combined capacitance may be found using the method by which the combined resistance of parallel-connected resistors is found:

$$\frac{1}{C_t} = \frac{1}{C_1} + \frac{1}{C_2} + \frac{1}{C_3}$$

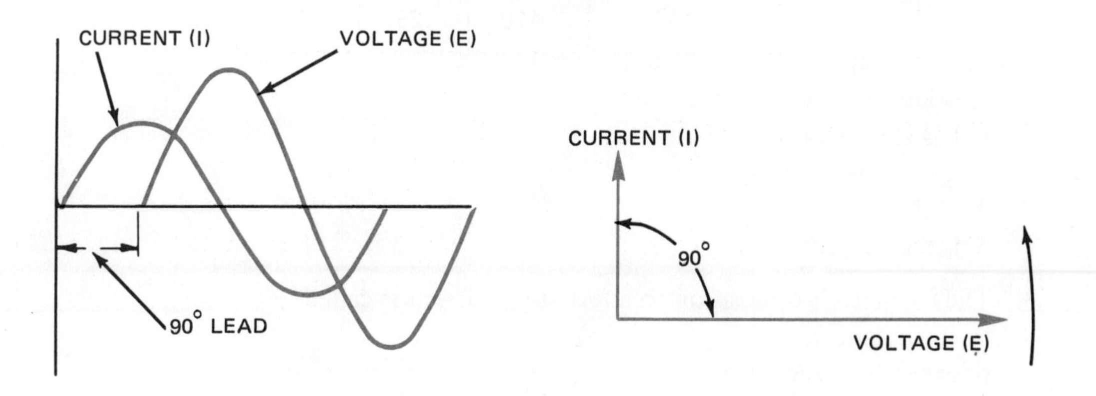

Fig. 3-2 Current leading voltage by 90 degrees.

Fig. 3-3 Vector diagram of current leading voltage by 90 degrees.

FINDING CURRENT

Figure 3-4 and the example show how the total current is determined in a capacitive ac circuit.

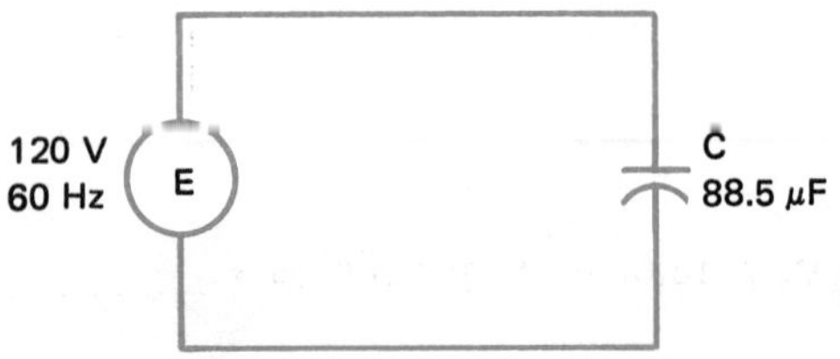

120 V
60 Hz
E
C
88.5 μF

Fig. 3-4 Finding current in a capacitive circuit.

First, find X_C:

$$X_C = \frac{1\,000\,000}{2\pi fC}$$

$$= \frac{1\,000\,000}{2 \times 3.14 \times 60 \times 88.5}$$

$$= \frac{1\,000\,000}{33\,300}$$

$$= 30\ \Omega$$

Using Ohm's Law:

$$I = \frac{E}{X_C} = \frac{120}{30}$$

$$= 4\ \text{A}$$

ACHIEVEMENT REVIEW

In the following problems, select the *best* answer to make the statement true. Place the letter of the selected answer in the space provided.

1. The most generally used unit of capacitive reactance is the ________
   a. farad.
   b. microfarad.
   c. ohm.
   d. henry.
   e. hertz.

2. Before a capacitor is handled, be sure it ________
   a. is large enough to do the job.
   b. has clean plates.
   c. has proper polarity.
   d. is charged.
   e. is discharged.

3. The capacitor's opposition to alternating current is called ________
   a. capacitance.
   b. capacitive reactance.
   c. resistance.
   d. the farad.
   e. the microfarad.

4. Capacitive reactance of a circuit may be increased by __________
   a. decreasing total capacitance.
   b. increasing total capacitance.
   c. increasing the number of farads.
   d. increasing source voltage.
   e. increasing frequency.

5. In a capacitive ac circuit __________
   a. current leads voltage.
   b. voltage leads current.
   c. current is in phase with voltage.
   d. current lags voltage.
   e. voltage is in phase with current.

6. The total capacitance of two 10-μF capacitors connected in parallel is __________
   a. 5 μF.
   b. 10 μF.
   c. 15 μF.
   d. 20 μF.
   e. 100 μF.

7. The total capacitive reactance of an ac circuit that draws 4 A from a 120-V, 60-Hz source is __________
   a. 2 Ω.
   b. 15 Ω.
   c. 30 Ω.
   d. 88.5 μF.
   e. 480 Ω.

8. The total capacitance of a 40-μF capacitor connected in series with a 60-μF capacitor is __________
   a. 24 μF.
   b. 40 μF.
   c. 60 μF.
   d. 100 μF.
   e. 2 400 μF.

9. The total capacitance, in μF, of an ac circuit that draws 0.362 A from a 120-V, 60-Hz source is __________
   a. 0.003.
   b. 8.
   c. 43.5.
   d. 166.
   e. 332.

10. The total current, in amperes, of an ac circuit that has a total capacitance of 638 μF connected to a 120-V, 25-Hz source is __________
    a. 4.8.
    b. 5.32.
    c. 8.0.
    d. 12.0.
    e. 25.5.

Series Circuit: Resistance and Inductance

OBJECTIVES

After studying this unit, the student will be able to

- explain the current-voltage relationship in an ac series circuit containing resistance and inductance.
- apply vectors to the analysis of an RL ac series circuit.

RESISTANCE AND INDUCTANCE IN SERIES

Resistors, coils, and capacitors may be connected in series in several combinations. This unit is concerned with a circuit containing a resistor connected in series with a coil.

Resistors and coils offer opposition to alternating current. The voltage and current in a resistor are in phase. In a coil, however, the current lags the voltage drop across the coil by 90 degrees. The combined opposition of resistors and coils is called *impedance* and is measured in ohms.

Circuit A: Series Circuit Containing Two Resistors

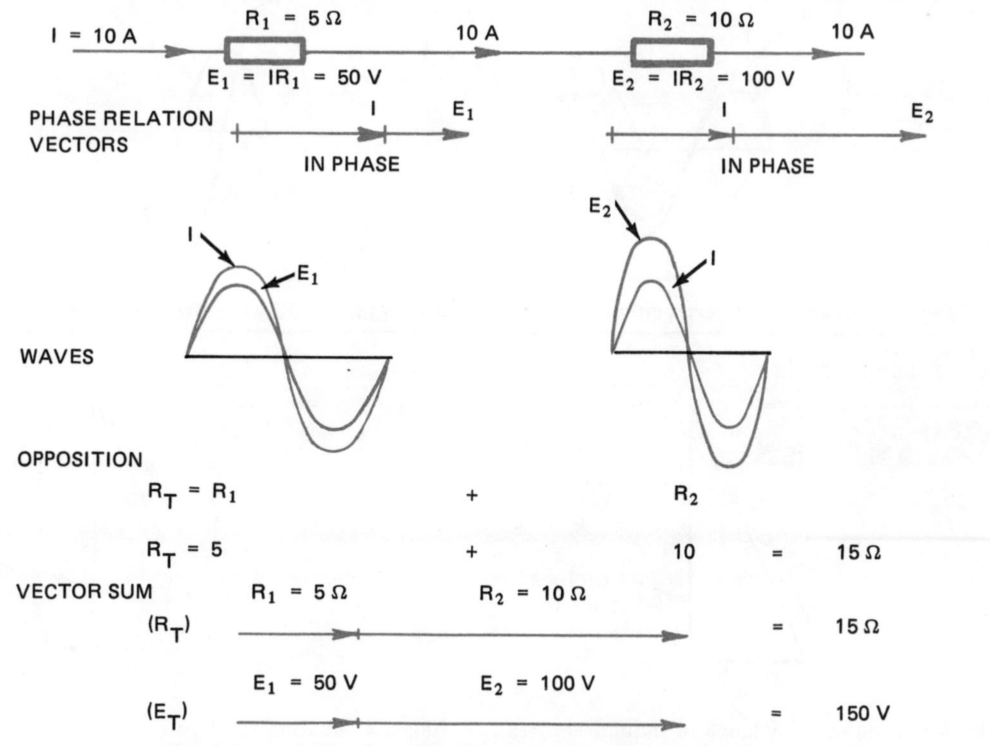

Fig. 4-1 Analysis of a series circuit containing two resistors.

The current in a series circuit is the same throughout; the total resistance is the sum of all the resistances in the circuit; and the total voltage is the sum of the voltage across each resistor, figure 4-1, page 21.

Circuit B: Series Circuit Containing a Resistor and a Coil

For the 5-ohm resistor in figure 4-2, the current is in phase with the voltage. The current in the coil lags the coil voltage by 90 degrees. The line current lags the total or line voltage by less than 90 degrees depending on the values of R and X_L. The resistance and the reactance must, therefore, be added vectorially to obtain the value of impedance.

The value of the impedance or total opposition, Z, can be calculated by substituting the values given in the formula in figure 4-3.

$$Z = \sqrt{5^2 + 10^2} = \sqrt{25 + 100} = \sqrt{125} = 11.2\ \Omega$$

Voltages E_R and E_L can be added vectorially as shown in figure 4-4 to obtain a total voltage E_T.

The calculation for total voltage is similar to the calculation for impedance.

$$E_T = \sqrt{50^2 + 100^2} = \sqrt{2\,500 + 10\,000} = \sqrt{12\,500} = 112\ V$$

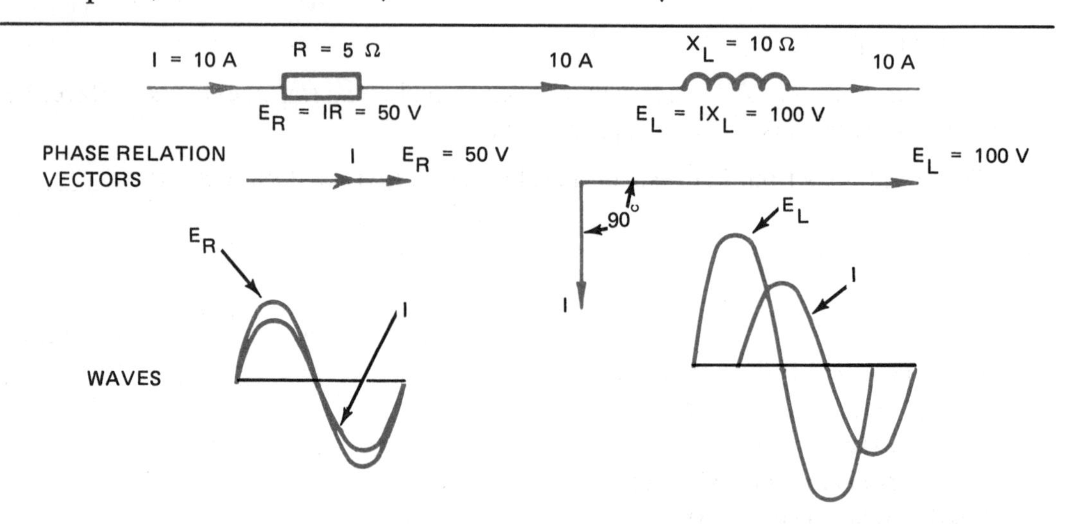

Fig. 4-2 Analysis of a series circuit containing a resistor and a coil with negligible resistance.

Impedance formula:

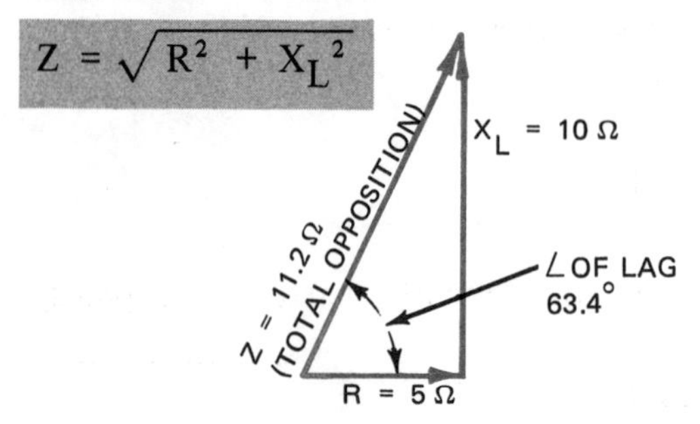

Fig. 4-3 Finding impedance of inductive series circuit.

$$E_T = \sqrt{E_R^2 + E_L^2}$$

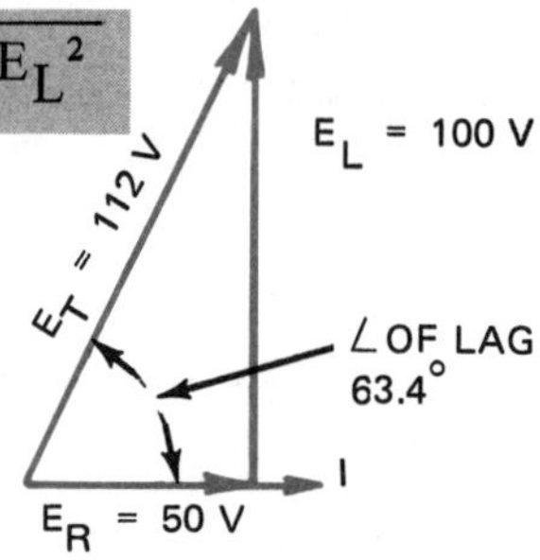

E_L = 100 V

E_T = 112 V

∠ OF LAG 63.4°

I

E_R = 50 V

Fig. 4-4 Finding total voltage for an inductive series circuit.

Notice that the current vector is included in figure 4-4. The current is in phase with E_R as expected. In an inductive circuit, the current always lags the voltage. In figure 4-4, the current lags the total voltage by 63.4°.

Phase Relationship. In figure 4-5, waveforms show the phase relationship of total current to total voltage.

Ohm's Law. As with all other types of circuits, Ohm's Law applies to this type of circuit.

$$I = \frac{E_T}{Z}, \quad E_T = IZ, \quad Z = \frac{E_T}{I}$$

Using the quantities given in the example in figure 4-4,

$$E_T = IZ$$
$$E_T = 10 \times 11.2 = 112 \text{ V}$$

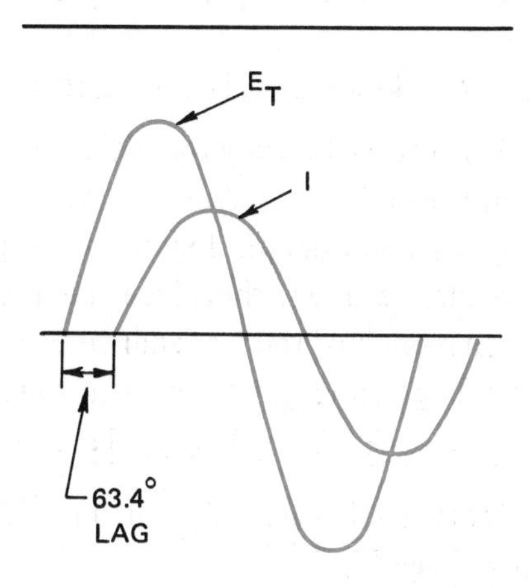

Fig. 4-5 Phase relationship.

ACHIEVEMENT REVIEW

In statements 1–7, select the *best* answer to make the statement true. Place the letter of the selected answer in the space provided.

1. The total opposition, in ohms, of a 3-ohm resistor with a 4-ohm coil is ________
   a. 3.
   b. 4.
   c. 5.
   d. 7.
   e. 12.

2. In an RL ac series circuit, the phase relationship of current through the coil to the voltage across the coil is that ________
   a. a 90° relationship exists.
   b. the voltage lags the current.
   c. there is always a 63.4° relationship.
   d. a 45° relationship exists.
   e. the waveforms are in phase.

3. An ac series circuit contains a 4-ohm resistor and an inductance coil. The voltage across the resistor is 80 volts, and the voltage across the coil is 60 volts. The total number of volts applied to the circuit is ________
   a. 20.
   b. 60.
   c. 80.
   d. 100.
   e. 140.

4. For the circuit described in problem 3, the total circuit current, in amperes, is ________
   a. 1.5.
   b. 2.
   c. 15.
   d. 20.
   e. 35.

5. For the circuit described in problem 3, state the angle between the voltage across the resistor and the total current without the use of a formula. ________
   a. 0°.
   b. 30°.
   c. 45°.
   d. 63.4°.
   e. 90°.

6. The total impedance, in ohms, of the circuit shown is ________ 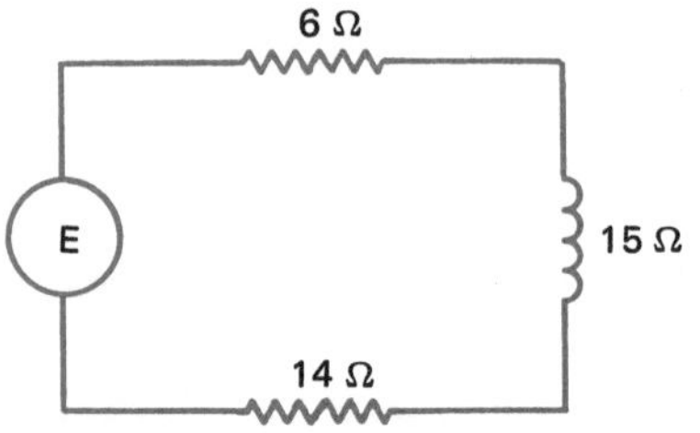
   a. 15.
   b. 20.
   c. 25.
   d. 29.
   e. 35.

7. A 40-ohm resistor is connected in series with a 0.079 6-H coil. The combination is connected to a 150-volt, 60-hertz source. The magnitude of total current, in amperes, is ________
   a. 2.14.
   b. 3.0.
   c. 3.75.
   d. 5.0.
   e. 30.

8. In the circuit shown, if E_R = 21 volts, and E_L = 28 volts, perform the following: 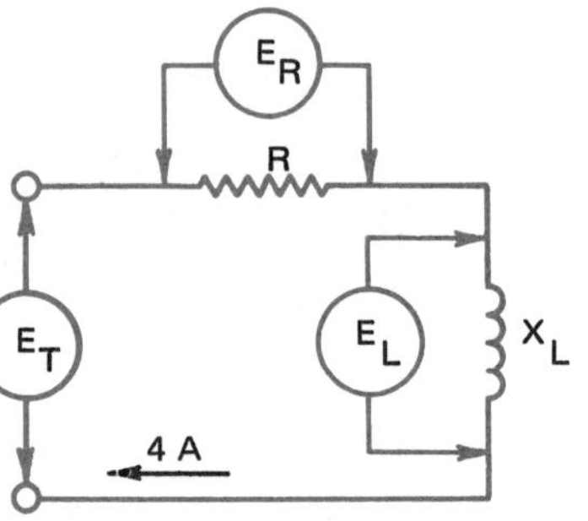
   a. Find the value of E_T.

   b. Find the value of R.

   c. Find the value of X_L.

   d. Draw the voltage vector diagram showing E_R, E_L, and E_T. Draw to scale.

5 Series Circuit: Resistance and Capacitance

OBJECTIVES

After studying this unit, the student will be able to

- describe the relationships of voltage and current in a series circuit containing resistance and capacitance.
- apply vectors in the analysis of an RC series circuit.

RESISTANCE AND CAPACITANCE IN SERIES

A combination frequently used in electrical circuits is a resistor in series with a capacitor. The total effect of this combination is similar to that of an inductive series circuit with the exception that the current leads the total voltage. The total opposition to current in RC circuits is called *impedance* (just as in RL circuits).

In the 5-ohm resistor in figure 5-1, the current is in phase with the voltage. The current in the capacitor leads the capacitor voltage by 90 degrees.

The line current leads the total or line voltage by less than 90 degrees depending on the values of R and X_C. The resistance and reactance must, therefore, be added vectorially to arrive at the value of impedance, figures 5-2 and 5-3, page 26.

The following calculation gives the same value of impedance as was obtained for the inductive series circuit in unit 4.

$$Z = \sqrt{R^2 + X_C^2} = \sqrt{5^2 + 10^2} = \sqrt{125} = 11.2\ \Omega$$

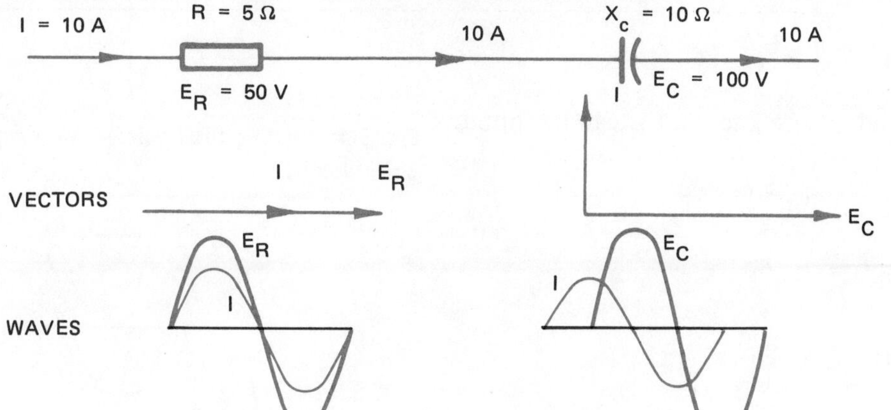

Fig. 5-1 Analysis of a series circuit containing a resistor and a capacitor.

In this case, however, the current leads the voltage by 63.4 degrees instead of lagging by the same amount. Compare figures 5-2 and 5-3. Note that the triangle is drawn below R instead of above it. X_L is in exact opposition to X_C.

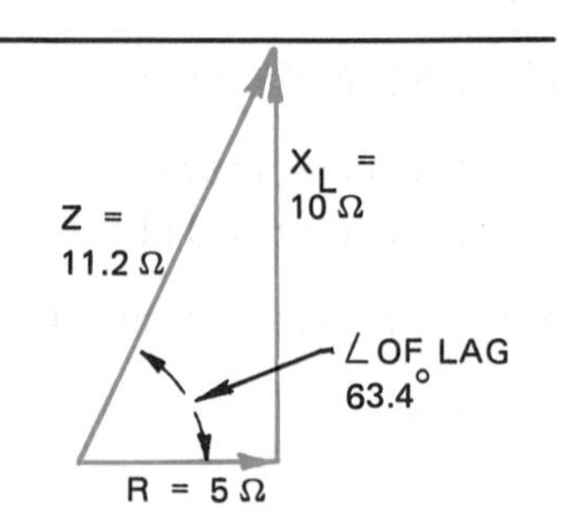

Fig. 5-2 Inductive series circuit.

Vector Sum of Voltages

As shown in figure 5-4, the voltages of each device in the circuit may be added vectorially as in the inductive series circuit. However, the vector diagram will be in the same position as the triangle in figure 5-3.

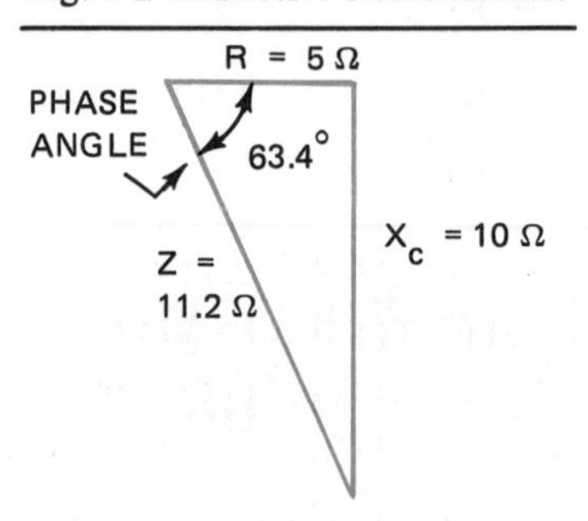

Fig. 5-3 Capacitive series circuit.

$$E_T = \sqrt{E_R^2 + E_C^2}$$
$$= \sqrt{50^2 + 100^2}$$
$$= 111.8 \text{ V}$$

Phase Relationship

In figure 5-5, the waveforms show that the current leads the total voltage by 63.4° (using the values in figure 5-4).

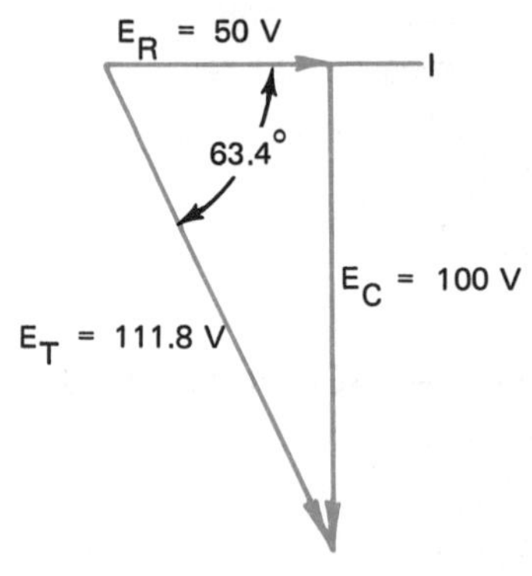

Fig. 5-4 Finding total voltage for a capacitive series circuit.

Summary of an RC series Circuit

$$Z = \sqrt{R^2 + X_C^2}$$

$$I_T = \frac{E_T}{Z}$$

$$E_T = \sqrt{E_R^2 + E_C^2}, \text{ or}$$

$$E_T^2 = E_R^2 + E_C^2$$

If E_T and E_C are known, E_R can be found:

$$E_R^2 + E_C^2 = E_T^2$$

$$E_R^2 = E_T^2 - E_C^2$$

$$E_R = \sqrt{E_T^2 - E_C^2}$$

Using the values in figure 5-4,

$$E_R = \sqrt{111.8^2 - 100^2} = 50 \text{ V}$$

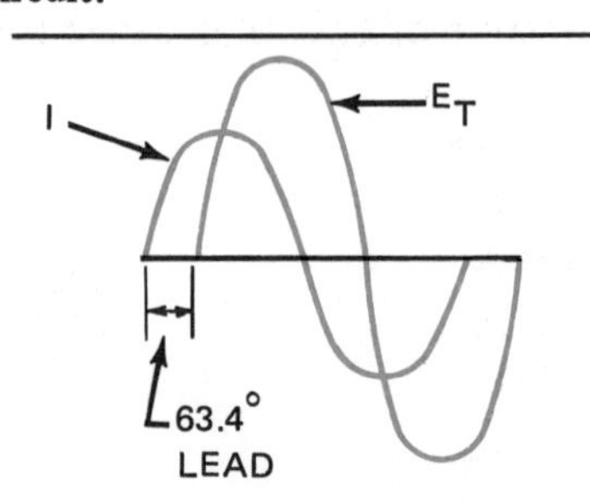

Fig. 5-5 Phase relationship.

ACHIEVEMENT REVIEW

Select the *best* answer for problems 1–7, and place the letter of the selected answer in the space provided.

1. The total current of an RC series ac circuit may be found with the expression ________

   a. $I = \frac{E_R}{R}$.
   b. $I = \frac{E_T}{X_c}$.
   c. $I = \frac{E_C}{R}$.
   d. $IR = E_T$.
   e. $I = \frac{E_R}{Z}$.

2. In an ac series RC circuit, ________
   a. current lags total voltage.
   b. resistor voltage leads current.
   c. the angle between total current and total voltage is 90°.
   d. the circuit phase angle is greater than 0° but less than 90°.
   e. the phase angle is 63.4°.

3. The total impedance, in ohms, of a series circuit containing two 30-ohm resistors and an 80-ohm capacitive reactance is ________
   a. 60.
   b. 80.
   c. 100.
   d. 110.
   e. 140.

4. In a series circuit, $E_R = 120$ volts, $E_C = 90$ volts, and $I = 3$ amperes. The total source voltage is ________
   a. 120 V.
   b. 150 V.
   c. 210 V.
   d. 270 V.
   e. 360 V.

5. If $E_T = 300$ volts, and $E_C = 180$ volts, the value of E_R in the RC series ac circuit is ________
   a. 0 V.
   b. 120 V.
   c. 180 V.
   d. 240 V.
   e. 480 V.

6. The value of the total current in the circuit shown is ________
   a. 1.71 A.
   b. 2.4 A.
   c. 3.0 A.
   d. 4.0 A.
   e. 12.0 A.

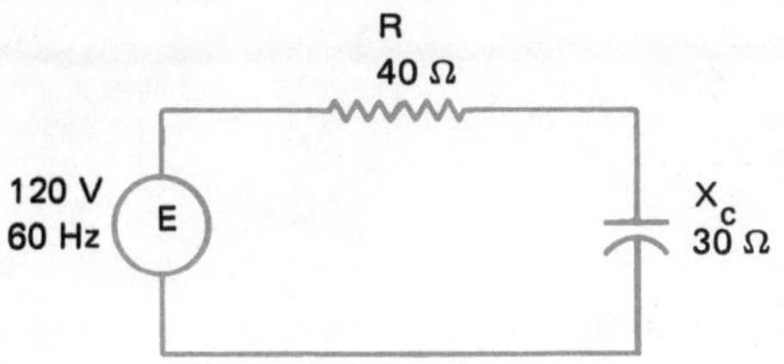

R
40 Ω
120 V
60 Hz
E
X_c
30 Ω

7. In the circuit shown, if the voltage across R_2 is 60 volts, the value of voltage across the capacitor is __________

   a. 40 V.
   b. 60 V.
   c. 90 V.
   d. 130 V.
   e. E.

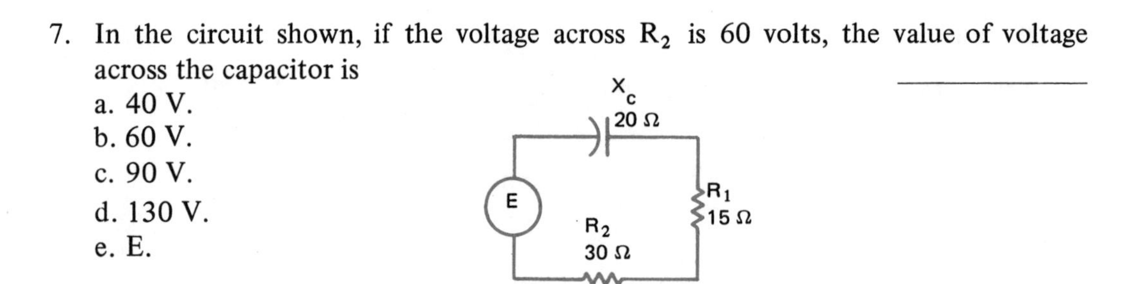

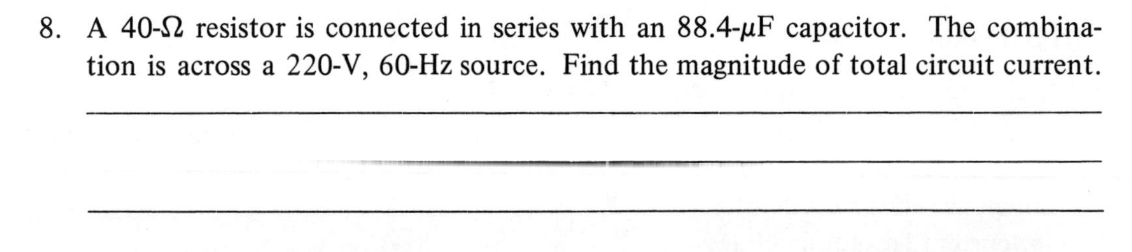

8. A 40-Ω resistor is connected in series with an 88.4-μF capacitor. The combination is across a 220-V, 60-Hz source. Find the magnitude of total circuit current.

   ____

   ____

   ____

   ____

Series Circuit: Resistance, Inductance, and Capacitance

OBJECTIVES

After studying this unit, the student will be able to

- discuss the effects of a combination of resistance, inductance, and capacitance connected in series.
- explain the relationships of voltage and current in a series RLC ac circuit.
- identify resonance in an ac series circuit.

RESISTANCE, INDUCTANCE, AND CAPACITANCE IN SERIES

An ac circuit may be inductive or capacitive. It may include a series combination of resistance, inductance, and capacitance. When the inductive reactance equals the capacitive reactance a condition called *resonance* exists. Since capacitive reactance opposes inductive reactance, they cancel each other and, thus, the resistance of the circuit is the only opposition.

Three series circuits with differing values of X_L, X_c and R are analyzed in this unit.

Circuit A: X_L is greater than X_c

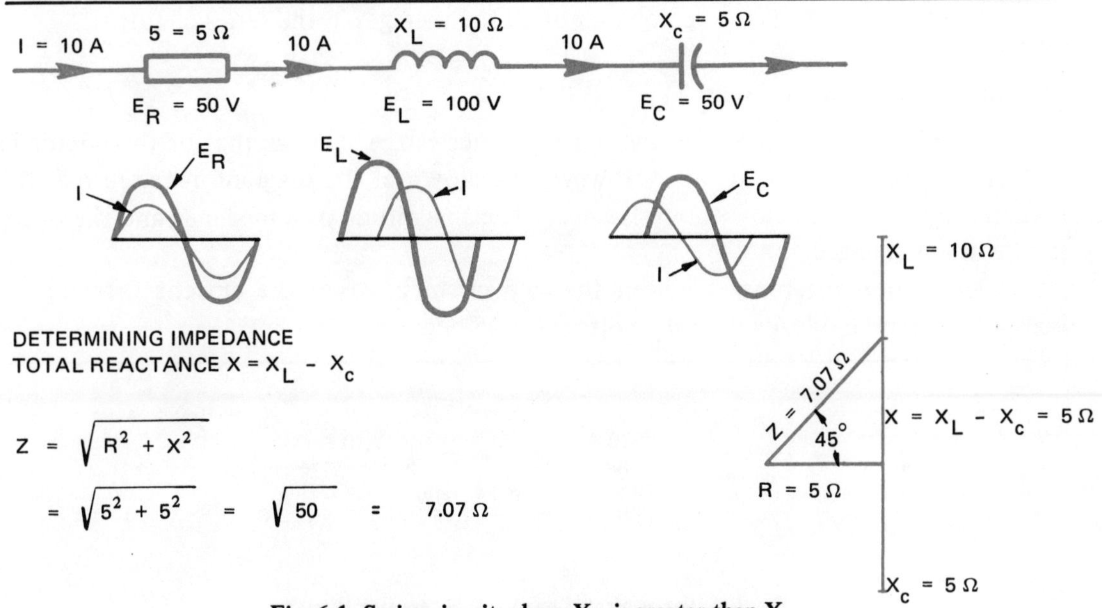

Fig. 6-1 Series circuit where X_L is greater than X_c.

With the quantities drawn to scale, X_L is twice as long as R.

The difference between a purely capacitive circuit as compared to a purely inductive circuit is that the capacitive effect leads by 90 degrees while the inductive effect lags by 90 degrees. Therefore, X_c is shown opposite to X_L in figure 6-2.

In the series RLC circuit, the resistor is common to both the capacitor and the inductor. The two diagrams shown in figure 6-2 can be combined as shown in figure 6-3. With X_L and X_c located directly opposite each other, the total reactance is equal to the difference of the two.

Since inductive reactance is greater than capacitive reactance, the net effect is that of an inductive circuit containing 5-ohms resistance and 5-ohms inductive reactance. The phase relationship between current and voltage is 45 degrees.

X_L = 10 Ω
R = 5 Ω
R = 5 Ω
X_C = 5 Ω

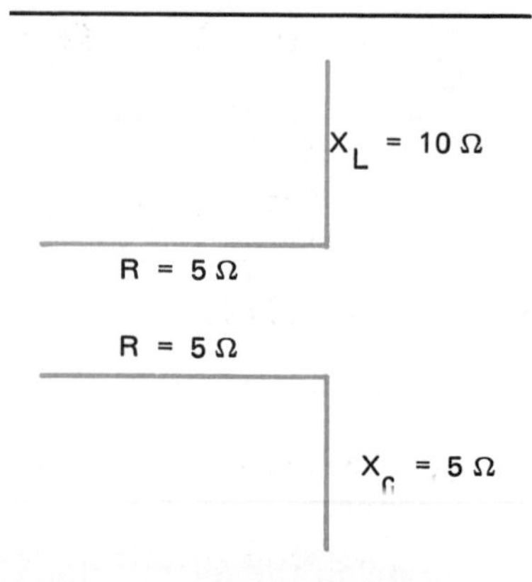

Fig. 6-2 X_L greater than X_c.

Vector Analysis to Obtain Total Voltage across Reactances

$$E_X = E_L - E_C = 50 \text{ V}$$

$$E_T = \sqrt{E_R^2 + E_X^2}$$

$$= \sqrt{50^2 + 50^2}$$

$$= \sqrt{5\,000}$$

$$= 70.7 \text{ V}$$

E_L = 100 V
E TOTAL = 70.7 V
45°
$E_X = E_L - E_C$ = 50 V
E_R = 50 V
I
E = 50 V

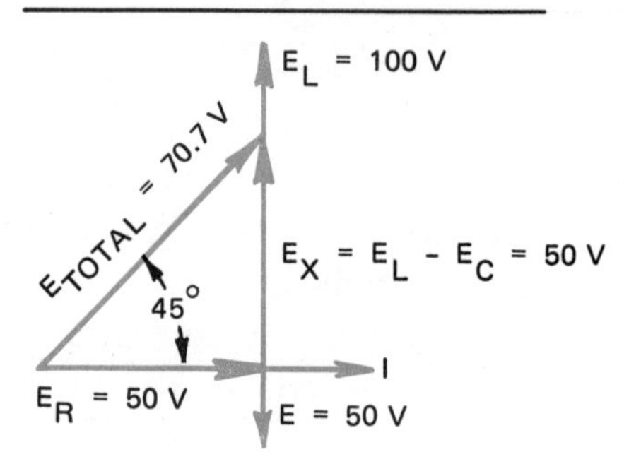

Fig. 6-3 X_L greater than X_c.

The total voltage is the vector sum of all the voltages in the series circuit.

Circuit B: X_c is Greater than X_L

In this circuit, figure 6-4, the net reactance is the same as that of the circuit in which X_L is greater than X_c. However, as shown in the diagram in figure 6-5, the reactance is located below the resistance. The magnitude of impedance and the phase angle are unchanged.

The major difference between the two circuits is that the current through the circuit in figure 6-4 leads the total voltage by 45 degrees.

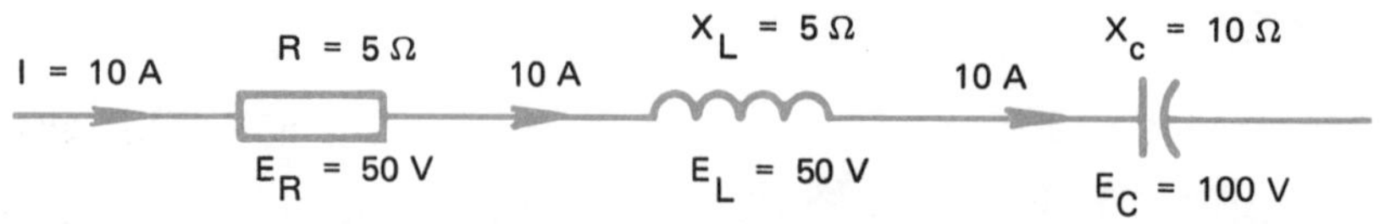

I = 10 A
R = 5 Ω
10 A
X_L = 5 Ω
10 A
X_c = 10 Ω
E_R = 50 V
E_L = 50 V
E_C = 100 V

Fig. 6-4 Series circuit where X_c is greater than X_L.

Determining Impedance

Total reactance $X = X_c - X_L$

$= 10 - 5 = 5\,\Omega$

$Z = \sqrt{R^2 + X^2}$

$= \sqrt{5^2 + 5^2} = \sqrt{50}$

$= 7.07\,\Omega$

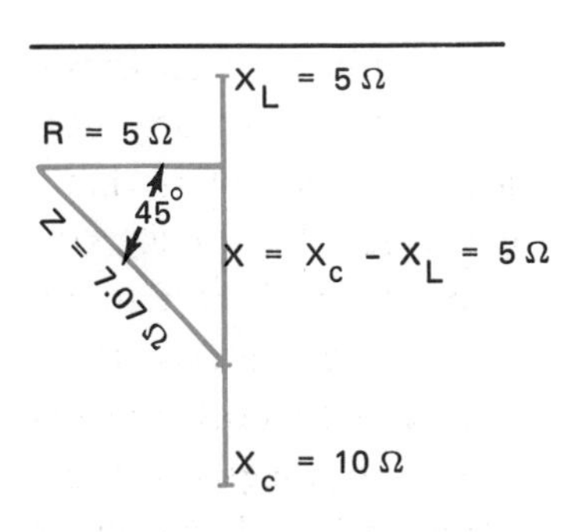

Fig. 6-5 X_c greater than X_L.

Vector Analysis to Determine Total Voltage

The total voltage is found in the same manner as in circuit A. The vector diagram for circuit B is similar to the impedance diagram for the circuit, figure 6-5, except that corresponding voltage vectors represent E_R, E_L, E_C, and E_T.

$E_X = E_C - E_L = 100 - 50 = 50\text{ V}$

$E_{total} = \sqrt{V_R^2 + V_X^2}$

$= \sqrt{50^2 + 50^2} = 70.7\text{ V}$

Circuit C: $X_L = X_c$ (Resonance)

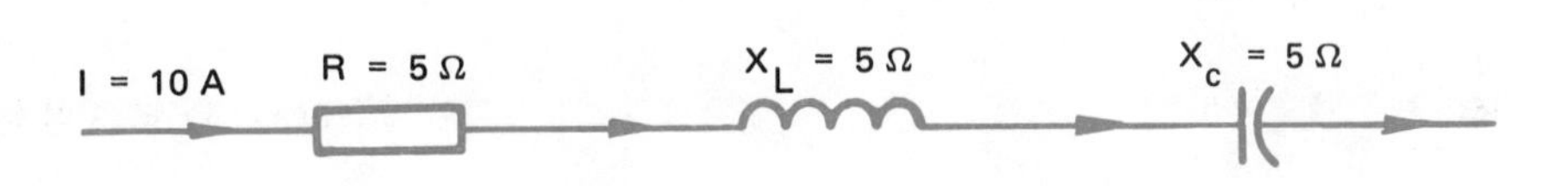

Fig. 6-6 Series circuit where X_L equals X_c.

Determining Impedance

$X = X_L - X_c = 0$

$Z = \sqrt{R^2 + 0} = R$

$= 5\,\Omega$

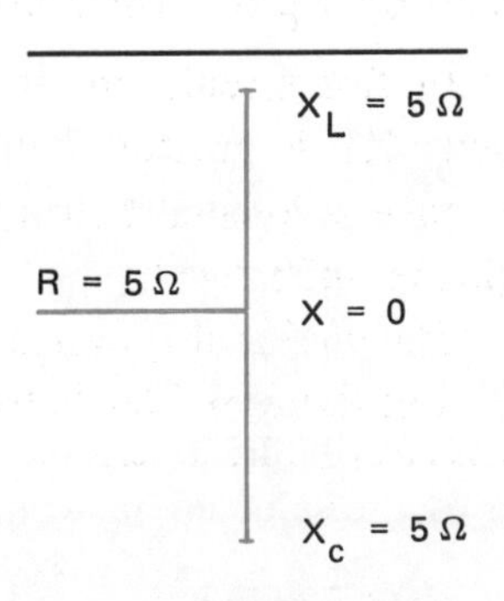

Fig. 6-7 X_L equal to X_c.

Total opposition or impedance in a resonant circuit is equal to or limited to the resistance. The angle between the total voltage and the current is zero. In other words, the current of the circuit is in phase with the total or line voltage.

Vector Analysis to Determine Total Voltage

$$E_X = E_L - E_C = 0$$

$$E_T = \sqrt{E_R^2 + 0} = E_R$$

$$= 50 \text{ V}$$

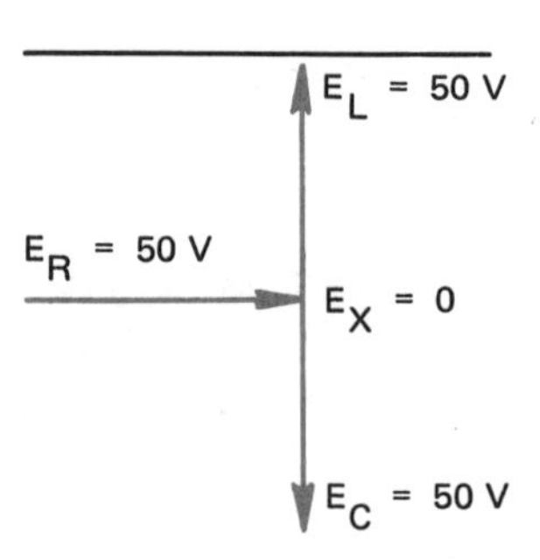

Fig. 6-8 E_L equal to E_C.

The voltage necessary to cause a current of 10 amperes to pass through this circuit is 50 volts instead of 70.7 volts as is required in the other two RLC series circuits.

Summary of Resonant Circuits

1. Opposition is at a minimum, Z = R

Circuit	R (ohms)	X_L (ohms)	X_c (ohms)	Z (ohms)
A	5	10	5	7.07
B	5	5	10	7.07
C (Resonance)	5	5	5	5

2. Maximum current exists at resonance, when impedance is at a minimum value. For example, if a source voltage of 70.7 volts is assumed for circuits A, B, and C:

$$I_A = \frac{E_T}{Z_A} = \frac{70.7}{7.07} = 10 \text{ A}$$

$$I_B = \frac{E_T}{Z_B} = \frac{70.7}{7.07} = 10 \text{ A}$$

$$I_C = \frac{E_T}{Z_C} = \frac{70.7}{5} = 14.14 \text{ A}$$

3. Phase relationship
   At resonance, the current is in phase with the total voltage.
4. Voltage relationship
   $E_T = E_R$

ACHIEVEMENT REVIEW

In problems 1–7, select the *best* answer to complete the statement. Place the letter of the selected answer in the space provided.

1. If X_c is greater than X_L in a series RLC circuit, the __________
   a. total current will lag the total voltage.
   b. total voltage is in phase with the current.
   c. current is in phase with E_C.
   d. current is in phase with E_L.
   e. current leads the total voltage.

2. A series ac circuit consists of R = 9 ohms, X_L = 20 ohms, and X_c = 8 ohms. The total reactance, in ohms, is __________
   a. 12. b. 15. c. 21. d. 28. e. 37.

3. An ac RLC series circuit has the following quantities: R = 40 ohms, X_c = 50 ohms, and X_L = 20 ohms. The total impedance, in ohms, is __________
   a. 30. b. 50. c. 70. d. 90. e. 110.

4. A series circuit has a resistor, a capacitor, and an inductor. A voltmeter is used to find the voltages of E_R = 50 volts, E_C = 70 volts, and E_L = 20 volts. The source voltage is equal to __________
   a. 20 volts.
   b. 50 volts.
   c. 70.7 volts.
   d. 100 volts.
   e. 140 volts.

5. A source voltage of an ac series RLC circuit is 120 volts. The circuit consists of the following quantities: R = 20 ohms, X_L = 40 ohms, and X_c = 40 ohms. The total circuit current, in amperes, is __________
   a. 1.2. b. 1.5. c. 2.0. d. 3.0. e. 6.0.

6. The circuit shown is a resonant circuit. If the total current is 10 amperes, the total applied voltage is equal to __________
   a. I X Z volts.
   b. 10 X Z volts.
   c. 20 volts.
   d. 200 volts.
   e. E.

R
20 Ω
E
X_c
X_L

7. In the circuit shown, the total current, in amperes, is __________
   a. 0.91.
   b. 1.67.
   c. 2.0.
   d. 3.33.
   e. 5.0.

75 Ω
250 V
E
150 Ω
50 Ω

8. In the circuit shown, find the value of voltage across X_c.

____

____

____

____

____

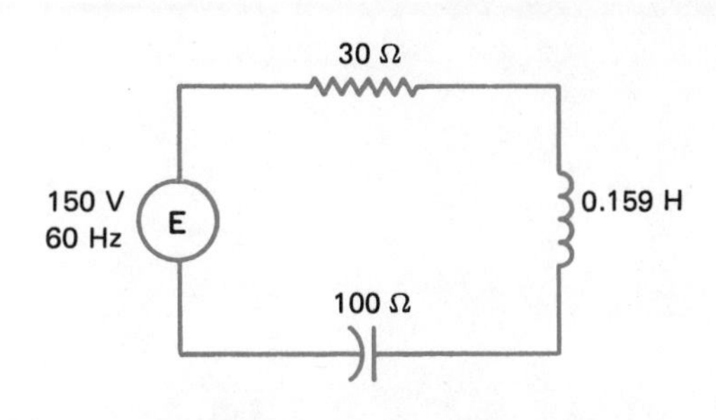

7 AC Parallel Circuits Containing Inductance

OBJECTIVES

After studying this unit, the student will be able to

- determine the current and voltage relationships in an alternating-current circuit containing a resistor connected in parallel with an inductor.
- make a vector analysis for an RL parallel circuit.

INDUCTANCE IN PARALLEL CIRCUITS

Parallel circuits are more common than series circuits because of the parallel or multiple system of energy transmission and distribution. It is not difficult to calculate the total current of a number of circuits connected in parallel. Remember that the current is the same in all parts of a series circuit. When the components of a series circuit are fixed values of resistance or reactance, the current is regarded as the reference point.

In a parallel circuit, each individual branch is connected directly across the line wires. Therefore, the voltage is the same across each branch of the circuit and, because of this, it is regarded as a fixed value or reference point in parallel circuit calculations.

Circuit A: Resistors in Parallel

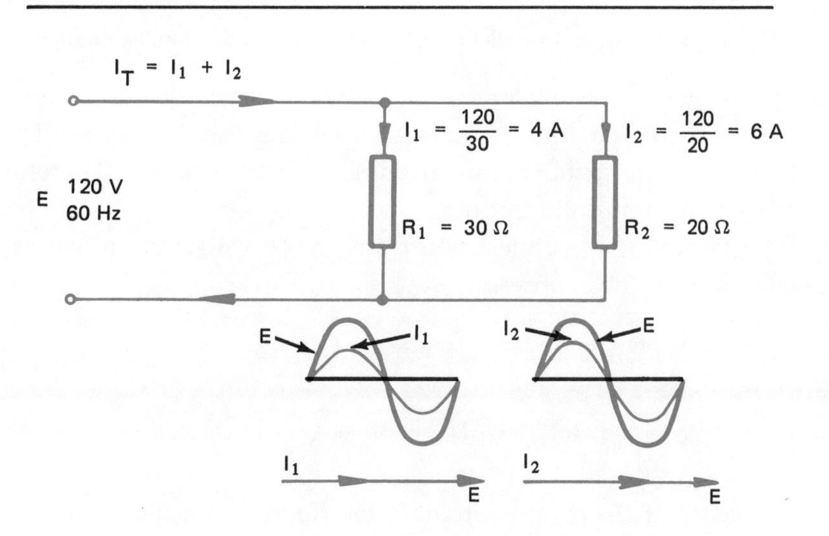

Fig. 7-1 Analysis of circuit containing resistors in parallel.

Both currents shown in figure 7-1, page 35, are in phase with the voltage. The total current may be determined by direct addition as in a dc circuit.

$$I_T = I_1 + I_2 = 10 \text{ A}$$

The total resistance can be calculated by dividing the total or line voltage by the line current.

$$R_T = \frac{E}{I_T} = \frac{120}{10} = 12\ \Omega$$

Circuit B: Resistor and Inductor in Parallel

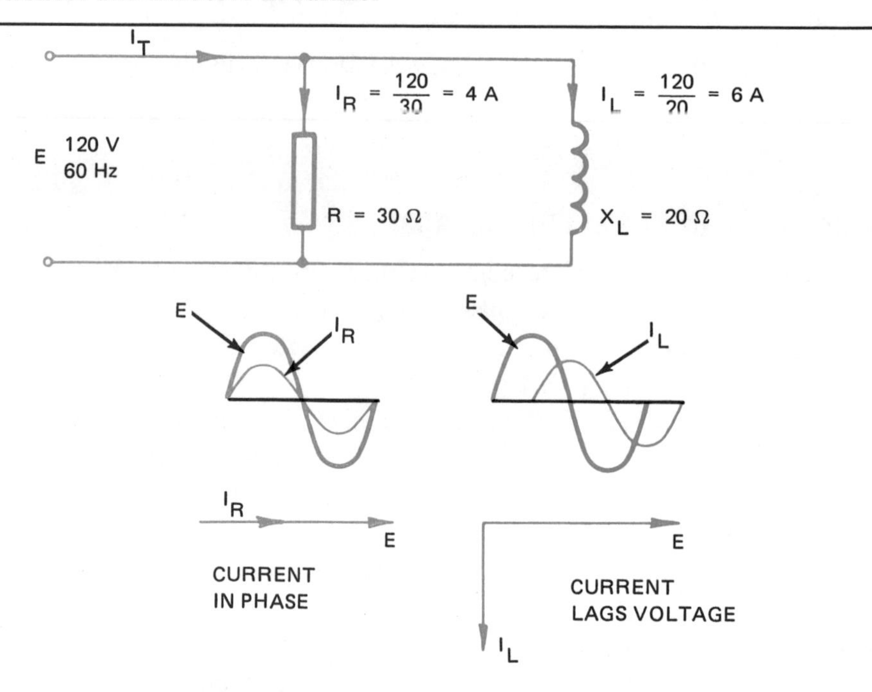

Fig. 7-2 Analysis of a parallel circuit containing a resistor and an inductor.

The current in the resistive branch of the parallel circuit is in phase with the line voltage, and the current in the inductive branch lags the line voltage by 90 degrees, figure 7-2. The two currents are out of phase with each other. Therefore, the total current is the vector sum of the two quantities.

The total current may be measured from a vector diagram drawn to scale, figure 7-3, or calculated with the expression:

$$I_T = \sqrt{I_R^2 + I_L^2}$$

For circuit B, figure 7-2,

$$I_T = \sqrt{4^2 + 6^2} = \sqrt{16 + 36} = \sqrt{52}$$

$$= 7.21 \text{ A}$$

The *impedance* of the parallel circuit is determined by using Ohm's Law:

$$Z = \frac{E}{I_T} = \frac{120}{7.21} = 16.6\ \Omega$$

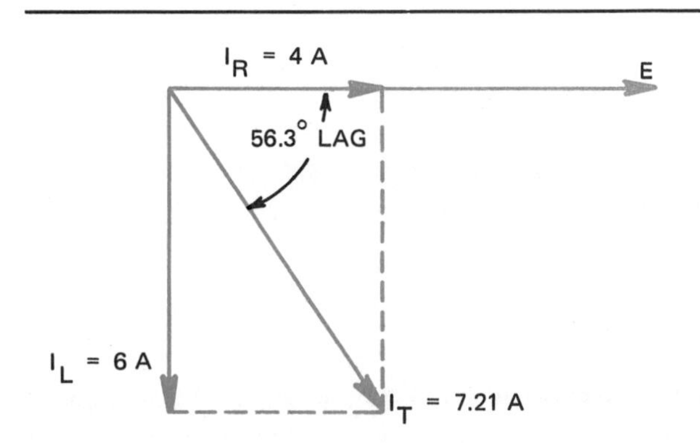

Fig. 7-3 Vector diagram of currents in a parallel circuit.

The phase angle lag of the total current may be measured on the vector diagram or it may be calculated. The value of this angle is less than 90 degrees. For circuit B, the phase angle is 56.3 degrees.

ACHIEVEMENT REVIEW

In problems 1–9, select the *best* answer and place the letter of the selected answer in the space provided.

1. In an RL parallel circuit, the opposition to total current is called __________
   a. reactance.
   b. resistance.
   c. phase relationship.
   d. a vector sum.
   e. impedance.

2. A 30-ohm resistor and a 40-ohm inductive reactance are connected in parallel to a 120-volt, 60-hertz source. The current through the reactance, in amperes, is __________
   a. 1.5.
   b. 1.71.
   c. 2.4.
   d. 3.0.
   e. 4.0.

3. An RL parallel circuit connected to a 120-V source has the following currents:
   I_R = 12 A, I_L = 16 A, I_T = 20 A.
   The total circuit impedance, in ohms, is __________
   a. 3.75.
   b. 4.28.
   c. 6.0.
   d. 7.5.
   e. 10.0.

4. A circuit consists of a resistor in parallel with an inductor. They are connected across a 120-V, 60-Hz source. The current through the resistor has a value of 8 A, and the current through the inductor has a value of 6 A. The value of the total current, in amperes, is __________
   a. 8.
   b. 10.
   c. 12.
   d. 14.
   e. 16.

5. In an RL parallel circuit, the ________
   a. total voltage and current are in phase.
   b. total voltage is in phase with the inductive current.
   c. inductive current is in phase with the resistive current.
   d. total current is 90° out of phase with the total voltage.
   e. total current and voltage are out of phase.

6. If a circuit contains a resistor in parallel with an inductive reactance, the ________
   a. voltage across the resistor is in phase with the voltage across the inductive reactance.
   b. total current lags the total voltage by 56.3°.
   c. total voltage lags the total current.
   d. total current and voltage are in phase.
   e. inductive current is in phase with the total voltage.

7. In a parallel RL circuit, the total voltage is ________
   a. in phase with the resistor current.
   b. in phase with the inductor current.
   c. in phase with the total current.
   d. out of phase with the total voltage by 56.3°.
   e. out of phase with the resistor current.

8. A 6-ohm resistor is connected in parallel to an 8-ohm inductive reactance and a 120-volt source. The value of the total current, in amperes, is ________
   a. 8.57.
   b. 15.0.
   c. 20.0.
   d. 25.0.
   e. 60.0.

9. A 6-ohm resistance is in parallel with a 4-ohm inductive reactance. If the resistive current is 18 amperes, the value of the total voltage is ________
   a. 36 volts.
   b. 72 volts.
   c. 108 volts.
   d. 180 volts.
   e. impossible to determine.

10. Find the magnitude of total impedance in the circuit shown.

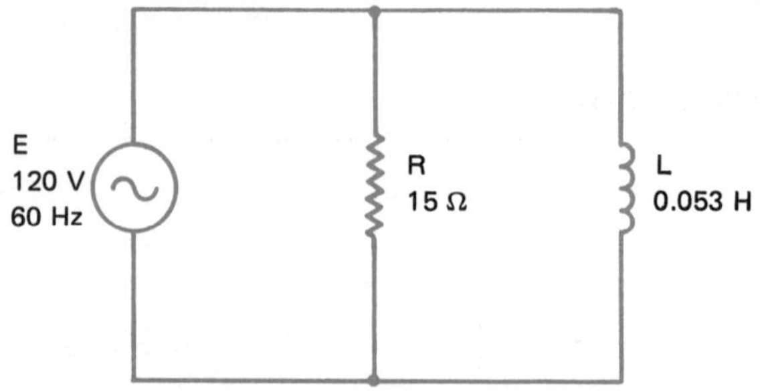

E
120 V
60 Hz
R
15 Ω
L
0.053 H

8 AC Parallel Circuits Containing Inductance and Capacitance

OBJECTIVES

After studying this unit, the student will be able to

- determine the current and voltage relationships in an alternating-current circuit containing resistance and capacitance in parallel.
- determine the current and voltage relationships in an alternating-current circuit containing resistance, inductance, and capacitance in parallel.
- discuss what is meant by antiresonance when applied to an alternating-current circuit containing resistance, inductive reactance, and capacitive reactance in parallel.

Most industrial power distribution lines have currents which lag the voltage. To offset this condition, power companies may connect banks of capacitors in parallel with the load. In this unit, the first circuit considered has a resistor connected in parallel with a capacitor. The second parallel circuit considered contains three branches: resistance, inductance, and capacitance. When the inductive current is equal to the capacitive current, a parallel resonance or antiresonance exists. This condition has a very practical application in the industrial field.

RESISTANCE AND CAPACITIVE REACTANCE IN PARALLEL

When a resistive load, such as a heating or lighting load, is connected in parallel with a capacitive reactance, certain voltage and current relationships result.

The current in the resistive branch is in phase with the line voltage. The current in the capacitive branch leads the line voltage by 90 degrees, figure 8-1, and figure 8-2, page 40.

Resistive Branch

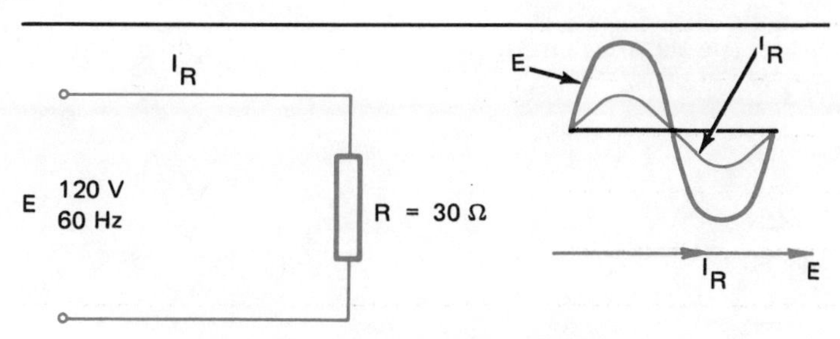

Fig. 8-1 Current in phase in resistive branch.

Capacitive Branch

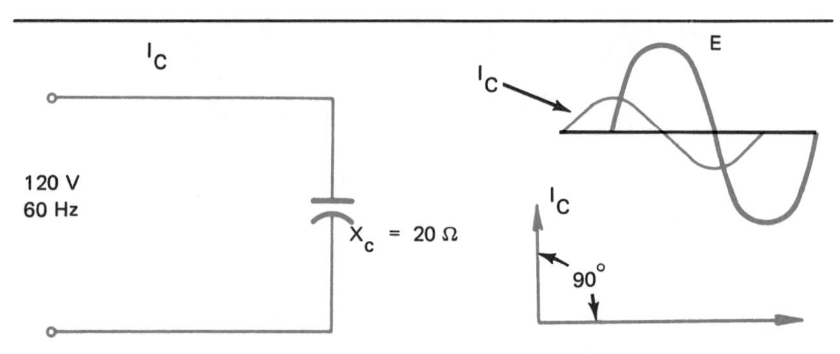

Fig. 8-2 Current leads the voltage in the capacitive branch.

Resistance and Capacitive Reactance in Parallel

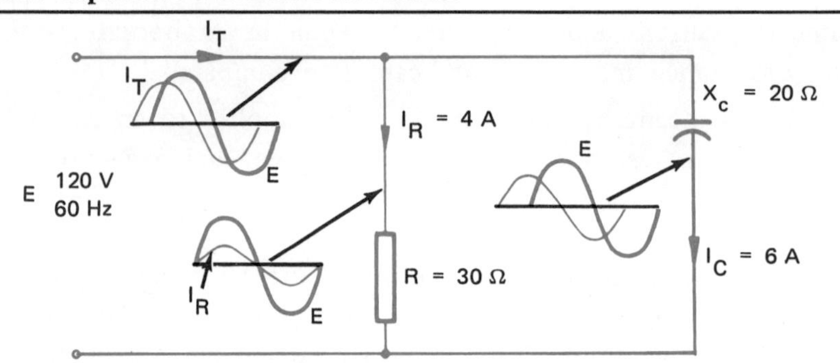

Fig. 8-3 Diagram of circuit containing a resistor connected in parallel with a capacitor.

When the resistive and capacitive branches are combined into a parallel circuit, as shown in figure 8-3, the total or line current is equal to the vector sum (not the arithmetic sum) of the currents taken by the individual branches. In other words, since the two currents are out of phase with each other, they must be added vectorially.

Solution of the Circuit (see figure 8-3):

$$I_R = \frac{E}{R}$$

$$I_R = \frac{120}{30} = 4 \text{ A}$$

Current in the capacitive reactance branch:

$$I_C = \frac{E}{X_c}$$

$$I_C = \frac{120}{20} = 6 \text{ A}$$

Total current in the parallel circuit:

$$I_T = \sqrt{I_R^2 + I_C^2}$$

$$I_T = \sqrt{4^2 + 6^2}$$

$$= \sqrt{52} = 7.21 \text{ A}$$

$$Z_T = \frac{E}{I_T} = \frac{120}{7.21} = 16.6\ \Omega$$

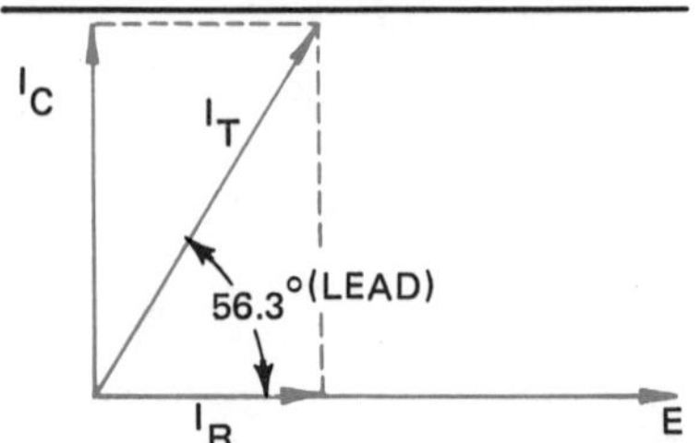

Fig. 8-4 Total current in the parallel RC circuit.

Circuit Containing Resistance, Inductive Reactance, and Capacitive Reactance in Parallel

With resistance, inductance, and capacitance connected in parallel, it is again necessary to find the current in each branch. To obtain the total or line current, the individual current values are added vectorially.

Circuit A. Parallel Circuit Having Unequal Reactances. The inductive current is greater than the capacitive current. Solution of the Circuit (see figure 8-5):

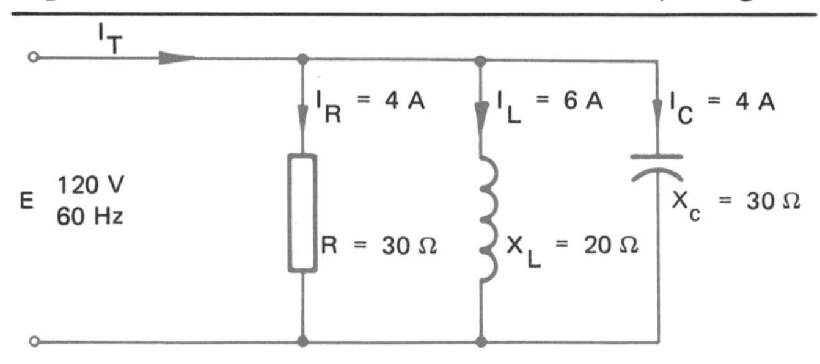

Fig. 8-5 Circuit diagram for parallel RLC circuit.

The current in the resistance branch is:

$$I_R = \frac{E}{R} = \frac{120}{30} = 4A$$

The current in the inductive branch is:

$$I_L = \frac{E}{X_L} = \frac{120}{20} = 6\text{ A}$$

The current in the capacitive branch is:

$$I_C = \frac{E}{X_C} = \frac{120}{30} = 4\text{ A}$$

The net reactive current is:

$$I_X = I_L - I_C = 6 - 4 = 2\text{ A}$$

Total current is equal to:

$$I_T = \sqrt{I_R^2 + I_X^2}$$

$$I_T = \sqrt{4^2 + 2^2}$$

$$= \sqrt{20} = 4.47\text{ A}$$

For parallel circuits having unequal reactances, where the inductive current is greater than the capacitive current, the following is true:

THE TOTAL CURRENT LAGS THE LINE VOLTAGE

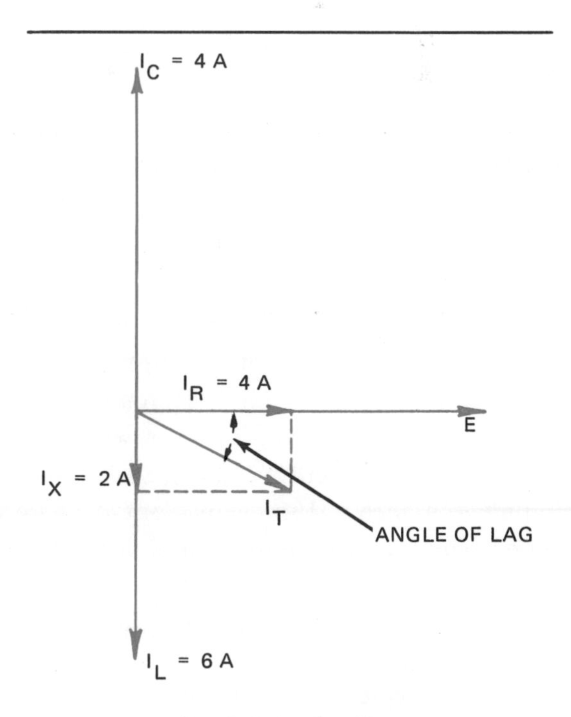

Fig. 8-6 Angle of lag.

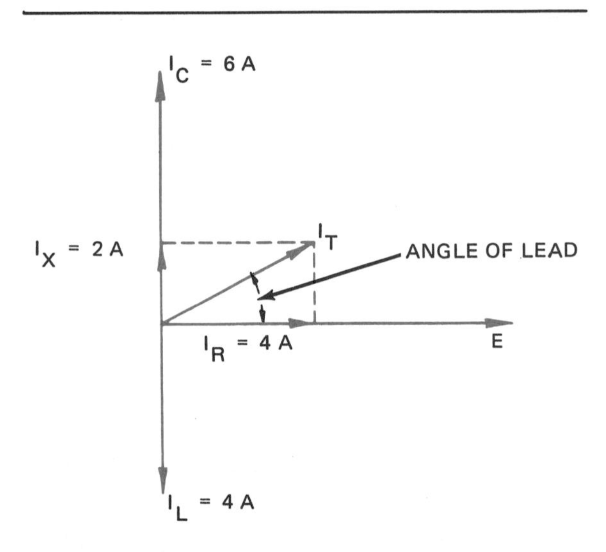

Fig. 8-7 Angle of lead.

Circuit B. Parallel Circuit Having Unequal Reactances. The capacitive current is greater than the inductive current. The vector analysis to determine the total current is shown in figure 8-7.

$$I_X = I_C - I_L = 6 - 4 = 2 \text{ A}$$

Total current is equal to:

$$I_T = \sqrt{I_R^2 + I_X^2}$$

$$I_T = \sqrt{4^2 + 2^2} = \sqrt{20} = 4.47 \text{ A}$$

For parallel circuits having unequal reactances, where the capacitive current is greater than the inductive current, the following is true:

THE TOTAL CURRENT LEADS THE LINE VOLTAGE

Circuit C. Resonance in a Parallel Circuit (Antiresonance). Resonance occurs in series circuits when the voltage across the inductance is equal to the voltage across the capacitance. The circuit current is then in phase with the line voltage and, for a given value of resistance, the current in the circuit is at a maximum value.

In a parallel circuit, however, when the current in the inductive branch equals the current in the capacitive branch, they cancel each other since the current through the inductive branch is 180 degrees out of phase with the current in the capacitive branch. As a result, total current or line current is the current through the resistive branch only. When this point is reached, the current is at a minimum value, figure 8-8. The circuit is said to be in antiresonance in contrast to resonance and maximum current in a series circuit.

Note: When I_L and I_C are equal, they cancel each other and the line current is at a minimum value. This is the antiresonance point.

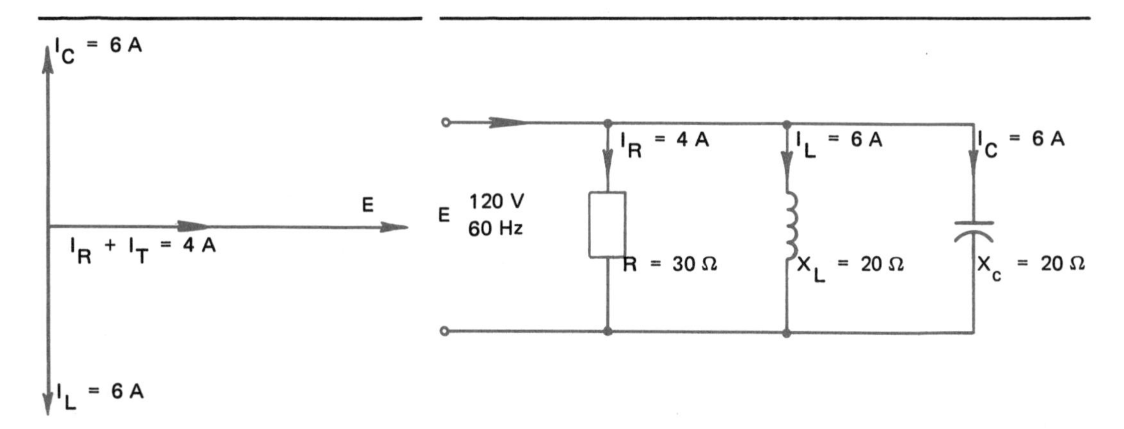

Fig. 8-8 Parallel resonance.

Fig. 8-9 Parallel resonant circuit.

Solution of the Circuit (see figure 8-9):

The current in the resistance branch is:

$$I_R = \frac{E}{R} = \frac{120}{30} = 4\ \text{A}$$

The current in the inductive reactance branch is:

$$I_L = \frac{E}{X_L} = \frac{120}{20} = 6\ \text{A}$$

The current in the capacitive reactance branch is:

$$I_C = \frac{E}{X_c} = \frac{120}{20} = 6\ \text{A}$$

Total line current is equal to:

$$I_T = \sqrt{I_R^2 + (I_L - I_C)^2}$$

When the circuit is in antiresonance:

$$I_L = \sqrt{I_R^2 + 0^2}$$

For the circuit in figure 8-9:

$$I_L = I_R = 4\ \text{A}$$

and

$$Z_{total} = \frac{E}{I_R} = R = 30\ \Omega$$

Summary of Parallel Resonant Circuits

The total or line current is equal to the current of the resistive branch and is at a minimum value.

Total impedance is equal to the resistance of the circuit.

The line current is in phase with the line voltage.

Power circuits carry out-of-phase currents such as those existing in all motor installations for the average industrial plant. These currents usually lag the voltage because of the inductive equipment. The values of these currents are much greater than the amount required for a given amount of power.

The desired conditions exist when the currents are in phase with the voltage. The lagging line current can be limited by connecting a capacitor or a bank of capacitors across the lines or in parallel with the particular equipment in question. This operation is called *power factor correction.*

ACHIEVEMENT REVIEW

In problems 1–8, select the *best* answer and place the letter of the selected answer in the space provided.

1. A circuit consists of a capacitor in parallel with a 0.1-henry inductor. The combination is across a 230-volt source. The capacitive current is 16 amperes and the inductive current is 12 amperes. The total line current, in amperes, is ________

   a. 1.2.
   b. 4.0.
   c. 20.
   d. 22.
   e. 28.

2. At antiresonance, the total ________
   a. impedance is equal to zero.
   b. current is equal to zero.
   c. current is at a maximum value.
   d. impedance is at a minimum value.
   e. current is in phase with the total voltage.

3. A 10-ohm resistor is in parallel with a 6-ohm inductive reactance. What value of capacitive reactance must be placed in parallel with the RL combination so that the total impedance is equal to 10 ohms? ________
   a. 3.75 ohms
   b. 6 ohms
   c. 10 ohms
   d. 16 ohms
   e. 60 ohms

4. A 15-ohm capacitive reactance is in parallel with a 20-ohm inductive reactance and a 150-volt, 60-hertz source. The value of the total current, in amperes, is ________
   a. 2.5.
   b. 6.0.
   c. 7.5.
   d. 10.
   e. 25.

5. An RLC parallel circuit has the following currents: I_R = 6 A, I_L = 10 A, I_C = 2 A. The value of total current, in amperes, is ________
   a. 4.
   b. 6.
   c. 8.
   d. 10.
   e. 18.

6. For the circuit shown, the total impedance value, in ohms, is ________
   a. 8.
   b. 10.
   c. 40.
   d. 50.
   e. 60.

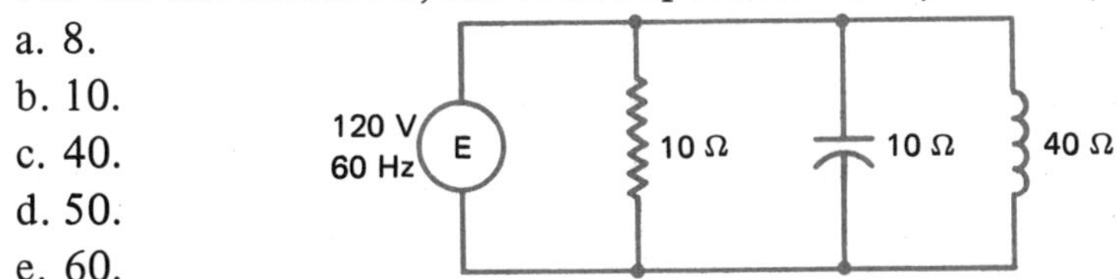

7. In an RLC parallel circuit, if X_L is greater than X_C, ________
   a. Z = R.
   b. total current lags total voltage.
   c. total current leads total voltage.
   d. total current and total voltage are in phase.
   e. total current is in phase with resistive current.

8. Which one of the listed answers is *not* a characteristic of parallel resonance? ________
   a. Line current is equal to the current of the resistive branch.
   b. Total impedance is equal to the value of resistance in the resistive branch.
   c. Line current is at a maximum value.
   d. Line current is in phase with the line voltage.
   e. The phase angle between line current and line voltage is zero degrees.

AC Power, Power Factor and Power Factor Correction

OBJECTIVES

After studying this unit, the student will be able to

- calculate the power in an ac circuit.
- discuss what is meant by power factor.
- explain the concept and importance of power factor correction.

The power in a dc circuit is equal to the product of voltage and current. In an ac circuit, the voltage and current are seldom in phase, except in incandescent lighting circuits and heating circuits. For most ac circuits then, the apparent power or product of voltage and current must be multiplied by a power factor to determine the true or real power.

POWER IN DC CIRCUITS

Power is the rate at which energy is used.

$$\text{Power [watts (W)]} = \text{volts} \times \text{amperes}$$

$$P = E \times I$$

$$1 \text{ kilowatt (kW)} = 1\,000 \text{ watts}$$

$$\text{kilowatts} = \frac{E \times I}{1\,000}$$

$$1 \text{ horsepower (hp)} = 746 \text{ watts}$$

$$\text{hp} = \frac{E \times I}{746}$$

$$\text{Percent Efficiency} = \frac{\text{Output Power}}{\text{Input Power}} \times 100$$

POWER IN AC RESISTIVE CIRCUITS

In a resistive-type alternating-current circuit with negligible inductance, such as circuits containing incandescent lights and heater loads, the power is determined in the same way as it is in direct-current circuits.

That is, $P = E \times I$

Power in the resistive circuit in figure 9-1 is $P = E \times I = 120 \times 12 = 1\,440$

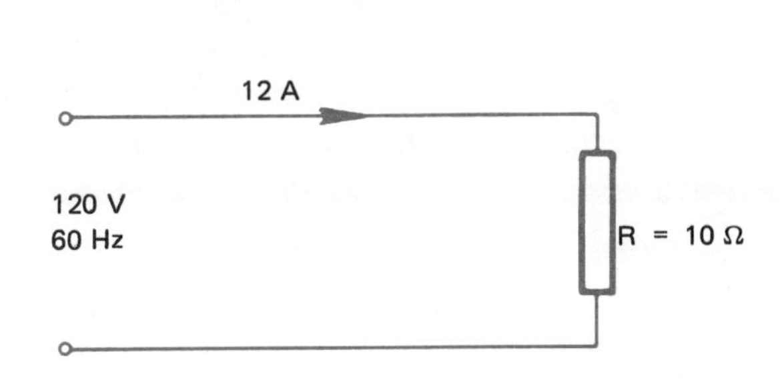

Fig. 9-1 Ac circuit resistance load.

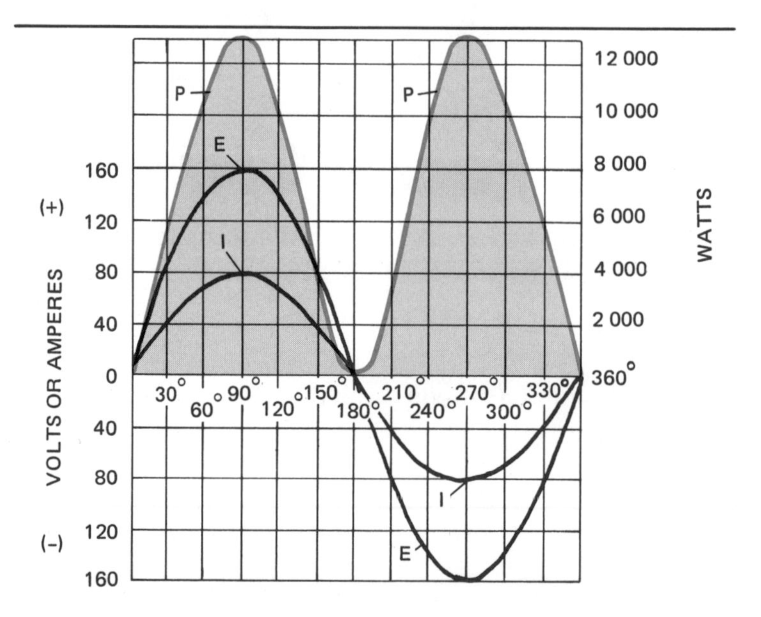

Fig. 9-2 Voltage, current, and power values for an ac resistive circuit.

watts. Examination of the voltage, current, and power cycles in figure 9-2 shows the in-phase relationship of current and voltage. During the first half of the cycle, the product of the positive instantaneous values of volts and amperes results in a positive cycle of power.

During the second half of the cycle, the power is the product of the negative voltage and current values. Since the product of any two negative values gives a plus quantity, the second cycle of power is also positive.

Fig. 9-3 Vector diagram of power in an ac resistive circuit.

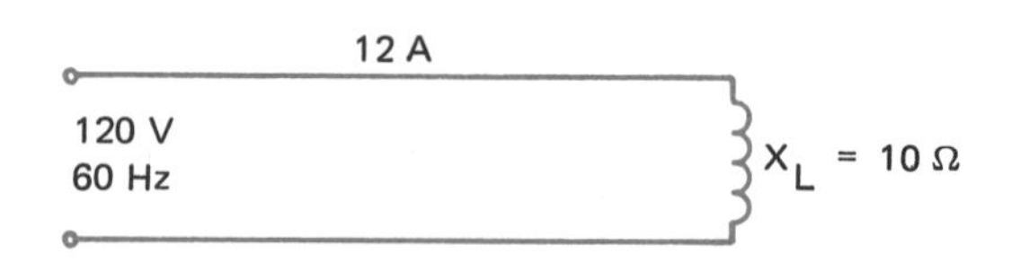

Fig. 9-4 Ac circuit inductive load.

POWER IN AC INDUCTIVE CIRCUITS

The current lags the voltage by 90 degrees in a circuit of pure inductance (no resistance), figure 9-5. The product of the instantaneous values of current and voltage during one cycle results in a power wave as shown in figure 9-5. During the first quarter of the voltage wave cycle, P = E × (–I) equals a negative power wave. During the second quarter of the cycle, the P wave is positive, P = E × (+I). During one voltage cycle two positive and two negative half cycles of power are produced.

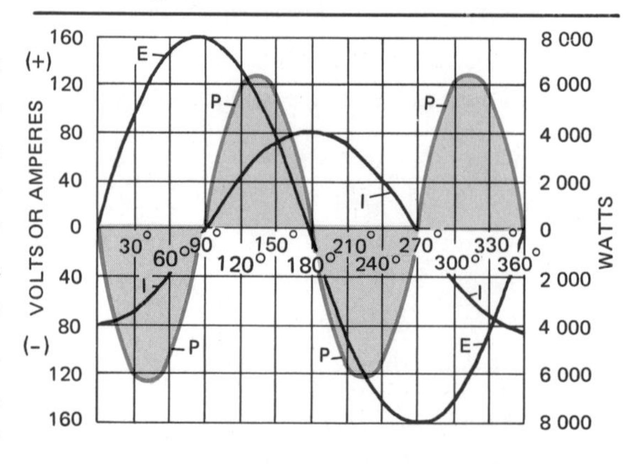

Fig. 9-5 Voltage, current, and power values for an ac inductive circuit.

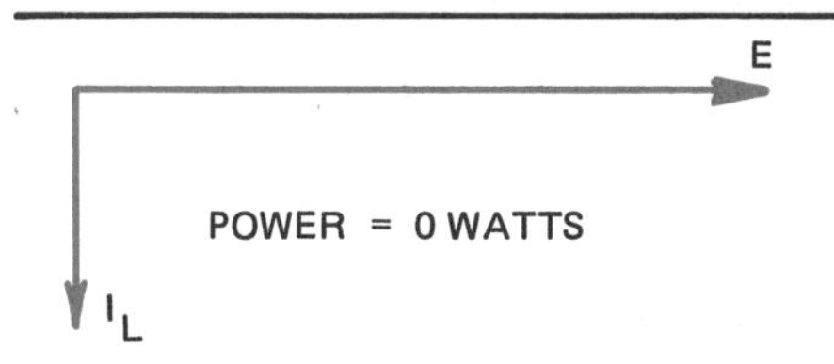

Fig. 9-6 Vector diagram of power in an ac inductive circuit.

As a result, the average power is zero. In the vector diagram, figure 9-6, no part of the current vector is in phase with the voltage. The in-phase portion of the inductive current is zero:

$$P = E \times (\text{in-phase portion of } I_L) = 120 \times 0 = 0$$

POWER IN AC CAPACITIVE CIRCUITS

As shown in figure 9-8, the current leads the voltage by 90 degrees. The average power is zero. The conditions are the same as in a pure inductive circuit, with the exception that the current leads the voltage by 90 degrees instead of lagging by the same amount.

POWER IN AC INDUCTIVE CIRCUITS CONTAINING RESISTANCE

Practically all motors, transformers, relays, and other inductive equipment have some resistance. As a result, the current lags the voltage by some angle less than 90 degrees, depending on the values of R and X_L. In figure 9-11, the shaded portion of the power wave is negative. This power portion increases as the phase angle displacement of current and voltage approaches 90 degrees.

At 90 degrees current lag, as in pure inductive and pure capacitive circuits, the net power is zero.

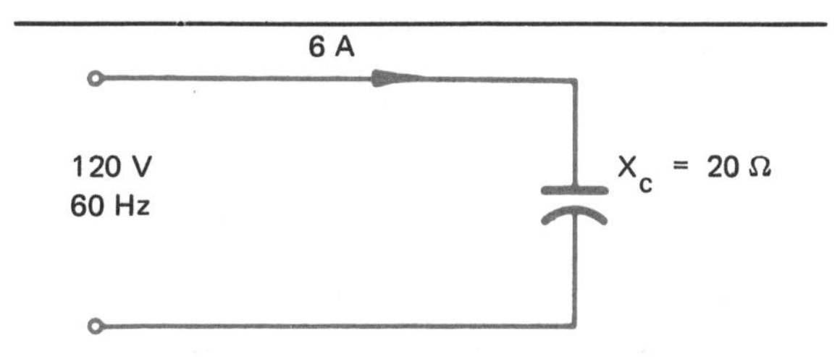

Fig. 9-7 Ac circuit capacitive load.

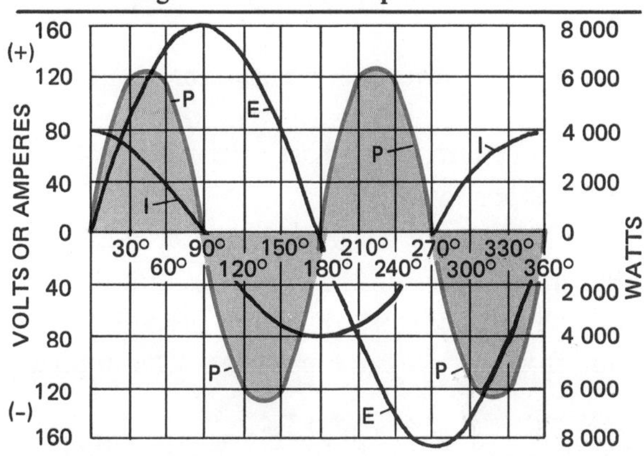

Fig. 9-8 Voltage, current, and power values for an ac capacitive circuit.

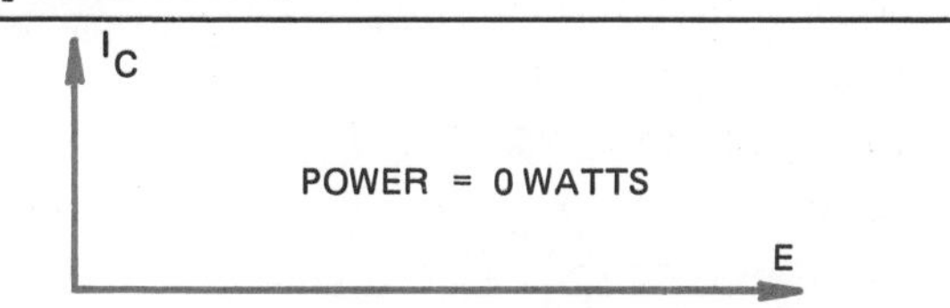

Fig. 9-9 Vector diagram of power in an ac capacitive circuit.

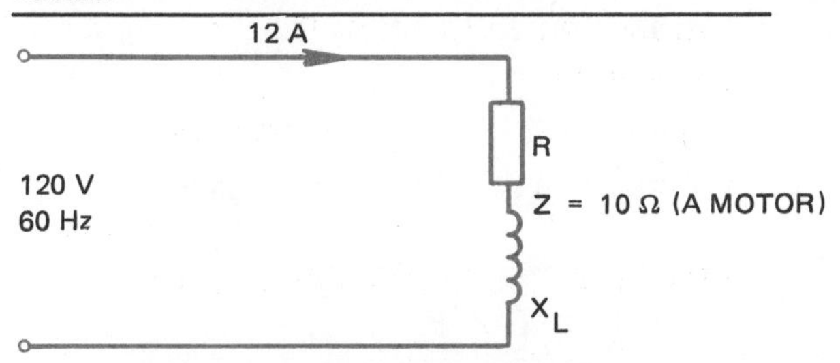

Fig. 9-10 Ac circuit impedance load.

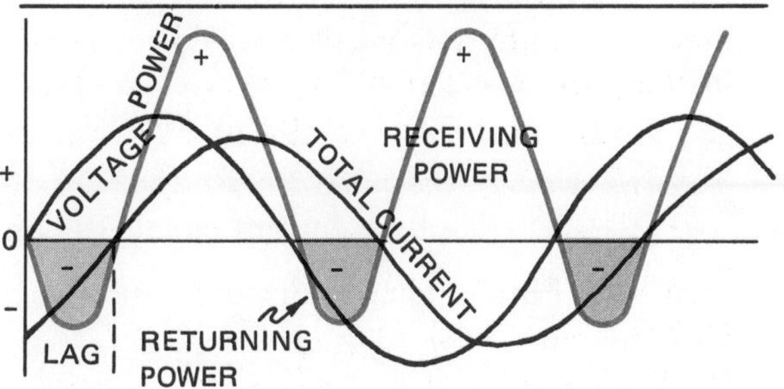

Fig. 9-11 Power in an ac inductive circuit containing resistance.

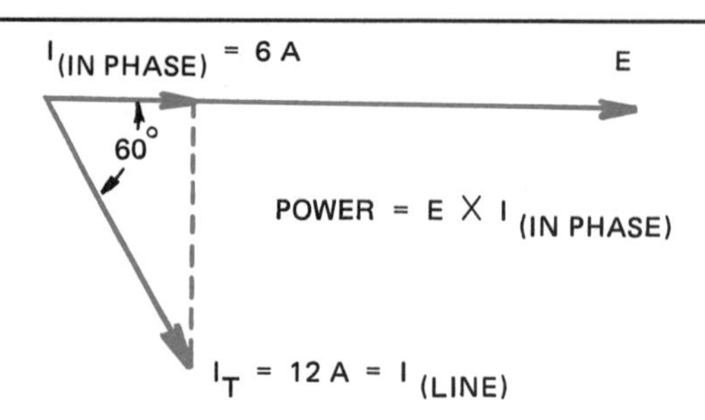

Fig. 9-12 Power in an ac circuit containing impedance: current lags the voltage by 60 degrees.

When current is in phase with the voltage, as in resistance circuits, the entire power wave is positive.

The vector diagram in figure 9-12 shows the actual line current passing through the impedance, such as a motor, transformer or relay. The part of the current that is actually doing work or aiding the voltage can be measured by drawing a perpendicular line from the 12-ampere vector. The in-phase current value is 6 amperes.

The ratio, $\frac{I_{(\text{in phase})}}{I_{\text{line}}}$ is called the cosine of the angle or the Power Factor (PF).

For figure 9-12, the power factor is equal to $\frac{I_{(\text{in phase})}}{I_{\text{line}}} = \frac{6}{12} = \frac{1}{2} = 0.5$

Power $= E \times I$ (in phase)

Since $\frac{I_{(\text{in phase})}}{I_{\text{line}}}$ = Power Factor,

then, $I_{(\text{in phase})} = I_{\text{line}} \times$ Power Factor.

Therefore, $P = E \times I_{\text{line}} \times$ Power Factor

$= 120 \times 12 \times 0.5 = 720$ watts.

POWER IN ALL CIRCUITS CONTAINING A SINGLE SOURCE

P = Voltage X Current X Power Factor

$= E \times I \times PF$

P (Watts) = Volts X Amperes X PF

Circuit A. Pure resistive circuit (Power Factor = 1)

P = 120 volts X 12 amperes X 1

= 1 440 watts

The in-phase current equals the line current.

Circuit B. Pure capacitive or pure inductive circuit (Power Factor = 0)

P = 120 volts X 12 amperes X 0

= 0

There is no in-phase current.

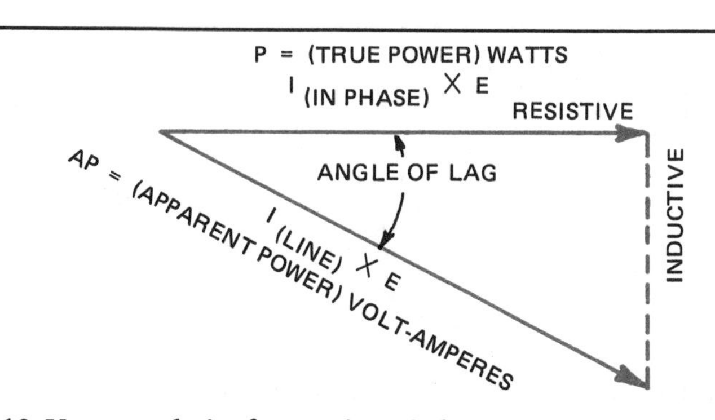

Fig. 9-13 Vector analysis of power in an inductive circuit having resistance.

POWER FACTOR

When all the current values shown in figure 9-12 are multiplied by the voltage, the result is a power diagram.

When line voltage and current are not in phase, the product of these two quantities is not power, but *apparent power*, figure 9-13.

AP = Apparent Power = $E_{line} \times I_{line}$ (Power that appears to be in the line)

The units of true power and apparent power can also be expressed in kilowatts (kW) and kilovolt-amperes (kVA).

True Power: $\frac{P}{1\,000}$ = kilowatts (kW)

$$\frac{\text{Power}}{1\,000} = \frac{E \times I_{\text{(in phase)}}}{1\,000} = \text{kW}$$

Apparent Power: $\frac{AP}{1\,000}$ = kilovolt-amperes (kVA)

$$\frac{\text{Apparent Power}}{1\,000} = \frac{E \times I_{\text{line}}}{1\,000} = \text{kVA}$$

The power factor can also be expressed in terms of apparent power.

$$P = \text{Power} = (E_{line} \times I_{line}) \times (PF)$$

$$= (AP) \times (PF)$$

$$\text{Power Factor} = \frac{\text{Power}}{\text{Apparent Power}}$$

$$PF = \frac{P}{AP} \text{ or } \frac{kW}{kVA}$$

Power Factor is sometimes specified in percent. To convert from a number to a percent, multiply the number by 100.

POWER FACTOR CORRECTION

The ideal power transmission occurs when the current is in phase with the voltage or when the power factor is 1. With this condition, a minimum amount of current is required to deliver a given amount of power. If line current can be reduced, less current need be generated to obtain the required results. This condition can be accomplished by connecting a capacitor or bank of capacitors across the line or in parallel with the inductive equipment, as shown in figure 9-14.

The size of the capacitor is determined from tables showing the relationship between the power factor, load kVA, and the kVA of the capacitors required for any desired new power factor. See figure 9-15.

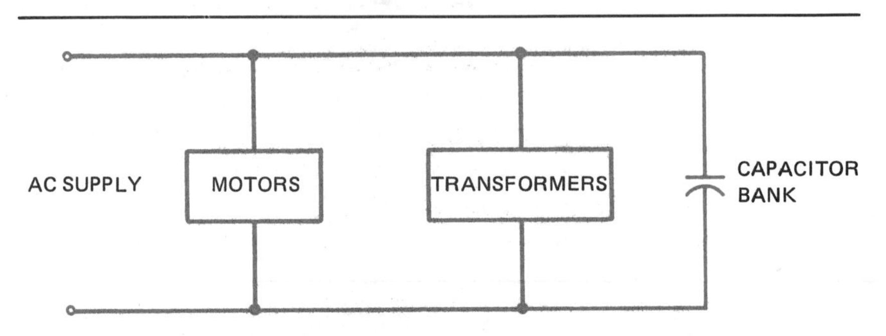

Fig. 9-14 Circuit showing power factor correction technique.

Power Factor – Correction Factor

Desired Power Factor in Percent

Original Power Factor in Percent	80	81	82	83	84	85	86	87	88	89	90	91	92	93	94	95	96	97	98	99	100
50	.982	1.008	1.034	1.060	1.086	1.112	1.139	1.165	1.192	1.220	1.248	1.276	1.303	1.337	1.369	1.403	1.441	1.481	1.529	1.590	1.732
51	.936	.962	.988	1.014	1.040	1.066	1.093	1.119	1.146	1.174	1.202	1.230	1.257	1.291	1.323	1.357	1.395	1.435	1.483	1.544	1.686
52	.894	.920	.946	.972	.998	1.024	1.051	1.077	1.104	1.132	1.160	1.188	1.215	1.249	1.281	1.315	1.353	1.393	1.441	1.502	1.644
53	.850	.876	.902	.928	.954	.980	1.007	1.033	1.060	1.088	1.116	1.144	1.171	1.205	1.237	1.271	1.309	1.349	1.397	1.458	1.600
54	.809	.835	.861	.887	.913	.939	.966	.992	1.019	1.047	1.075	1.103	1.130	1.164	1.196	1.230	1.268	1.308	1.356	1.417	1.559
55	.769	.795	.821	.847	.873	.899	.926	.952	.979	1.007	1.035	1.063	1.090	1.124	1.156	1.190	1.228	1.268	1.316	1.377	1.519
56	.730	.756	.782	.808	.834	.860	.887	.913	.940	.968	.996	1.024	1.051	1.085	1.117	1.151	1.189	1.229	1.277	1.338	1.480
57	.692	.718	.744	.770	.796	.822	.849	.875	.902	.930	.958	.986	1.013	1.047	1.079	1.113	1.151	1.191	1.239	1.300	1.442
58	.655	.681	.707	.733	.759	.785	.812	.838	.865	.893	.921	.949	.976	1.010	1.042	1.076	1.114	1.154	1.202	1.263	1.405
59	.618	.644	.670	.696	.722	.748	.775	.801	.828	.856	.884	.912	.939	.973	1.005	1.039	1.077	1.117	1.165	1.226	1.368
60	.584	.610	.636	.662	.688	.714	.741	.767	.794	.822	.850	.878	.905	.939	.971	1.005	1.043	1.083	1.311	1.192	1.334
61	.549	.575	.601	.627	.653	.679	.706	.732	.759	.787	.815	.843	.870	.904	.936	.970	1.008	1.048	1.096	1.157	1.299
62	.515	.541	.567	.593	.619	.645	.672	.698	.725	.753	.781	.809	.836	.870	.902	.936	.974	1.014	1.062	1.123	1.265
63	.483	.509	.535	.561	.587	.613	.640	.666	.693	.721	.749	.777	.804	.838	.870	.904	.942	.982	1.030	1.091	1.233
64	.450	.476	.502	.528	.554	.580	.607	.633	.660	.688	.716	.744	.771	.805	.837	.871	.909	.949	.997	1.058	1.200
65	.419	.445	.471	.497	.523	.549	.576	.602	.629	.657	.685	.713	.740	.774	.806	.840	.878	.918	.966	1.027	1.619
66	.388	.414	.440	.466	.492	.518	.545	.571	.598	.626	.554	.682	.709	.743	.775	.809	.847	.887	.935	.996	1.138
67	.358	.384	.410	.436	.462	.488	.515	.541	.568	.596	.624	.652	.679	.713	.745	.779	.817	.857	.905	.966	1.108
68	.329	.355	.381	.407	.433	.459	.486	.512	.539	.567	.595	.623	.650	.684	.716	.750	.788	.828	.876	.937	1.079
69	.299	.325	.351	.377	.403	.429	.456	.482	.509	.537	.565	.593	.620	.654	.866	.720	.758	.798	.840	.907	1.049
70	.270	.296	.322	.348	.374	.400	.427	.453	.480	.508	.536	.564	.591	.625	.657	.691	.729	.769	.811	.878	1.020
71	.242	.268	.294	.320	.346	.372	.399	.425	.452	.480	.508	.536	.563	.597	.629	.663	.501	.741	.783	.850	.992
72	.213	.239	.265	.291	.317	.343	.370	.396	.423	.451	.479	.507	.534	.568	.600	.634	.672	.712	.754	.821	.963
73	.186	.212	.238	.264	.290	.316	.343	.369	.396	.424	.452	.480	.507	.541	.573	.607	.645	.685	.727	.794	.936
74	.159	.185	.211	.237	.263	.289	.316	.342	.369	.397	.425	.453	.480	.514	.546	.580	.618	.658	.700	.767	.909
75	.132	.158	.184	.210	.236	.262	.289	.315	.342	.370	.398	.426	.453	.487	.519	.553	.591	.631	.673	.740	.882
76	.105	.131	.157	.183	.209	.235	.262	.288	.315	.343	.371	.399	.426	.460	.492	.526	.564	.604	.652	.713	.855
77	.079	.105	.131	.157	.183	.209	.236	.262	.289	.317	.345	.373	.400	.434	.466	.500	.538	.578	.620	.687	.829
78	.053	.079	.105	.131	.157	.183	.210	.236	.263	.291	.319	.347	.374	.408	.440	.474	.512	.552	.594	.661	.803
79	.026	.052	.078	.104	.130	.156	.183	.209	.236	.264	.292	.320	.347	.381	.413	.447	.485	.525	.567	.634	.776
80	.000	.026	.052	.078	.104	.130	.157	.183	.210	.238	.266	.294	.321	.355	.387	.421	.459	.499	.541	.608	.750
81	.....	.000	.026	.052	.078	.104	.131	.157	.184	.212	.240	.268	.295	.329	.361	.395	.433	.473	.515	.582	.724
82	.....	....	.000	.026	.052	.078	.105	.131	.158	.186	.214	.242	.269	.303	.335	.369	.407	.447	.489	.556	.698
83	....	....	....	.000	.026	.052	.079	.105	.132	.160	.188	.216	.243	.277	.309	.343	.381	.421	.463	.530	.672
84	....	....	....	....	.000	.026	.053	.079	.106	.134	.162	.190	.217	.251	.283	.317	.355	.395	.437	.504	.645
85	.....	....	....	....	....	.000	.027	.053	.080	.108	.136	.164	.191	.225	.257	.291	.329	.369	.417	.478	.620

Fig. 9-15.

The size of capacitor required to raise the power factor of a given load to a higher value can be found as follows:

Assume a 500-kVA load at a power factor of 0.6 or 60 percent: 500 kVA × 0.6 = 300 kW.

It is desired to raise the power factor to 90 percent. The capacitor kVA value required to accomplish this is found by multiplying 300 kW by the factor taken from the table in figure 9-15, or 0.85. (Locate the original power factor of 60 in the first column of the table on the left, and locate the desired power factor of 90 at the top of the table. Go to the right from 60 and down from 90 until the columns meet. This one value, 0.85, is the correction factor to be used in the calculation.) The capacitor required has a capability of 300 × 0.85 = 255 kVA.

Assume that the next higher standard capacitor rating is selected (300 kVA). What is the value of the resulting correction factor?

$$\frac{300}{300} = 1.00$$

Referring to figure 9-15, and using the original power factor of 60 percent and a correction factor of 1.00, the ultimate power factor is nearly 95 percent.

MEASUREMENT OF POWER

Using a Wattmeter

Power in ac circuits is measured with a *wattmeter*, figure 9-16. The method of connecting the wattmeter into a circuit to measure power is shown in figure 9-17.

Fig. 9-16 A portable wattmeter. (*Courtesy Weston Instruments, Inc.*)

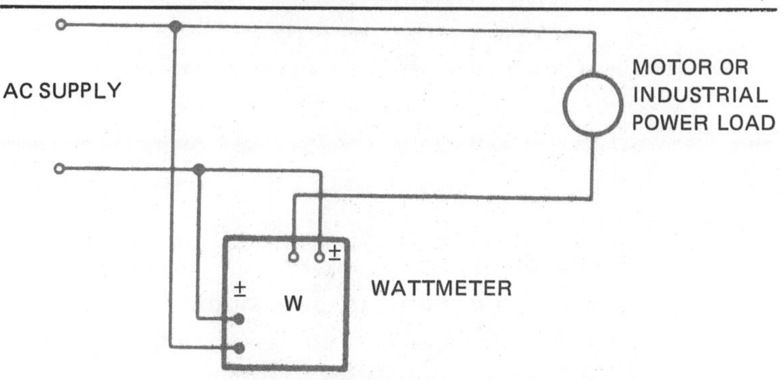

Fig. 9-17 Measurement of power using a wattmeter.

MEASUREMENT OF POWER FACTOR

Using a Power Factor Meter

The exterior of the power factor meter is identical to that of the wattmeter. The connections, shown in figure 9-18, are identical to those shown for the measurement of power in an ac line.

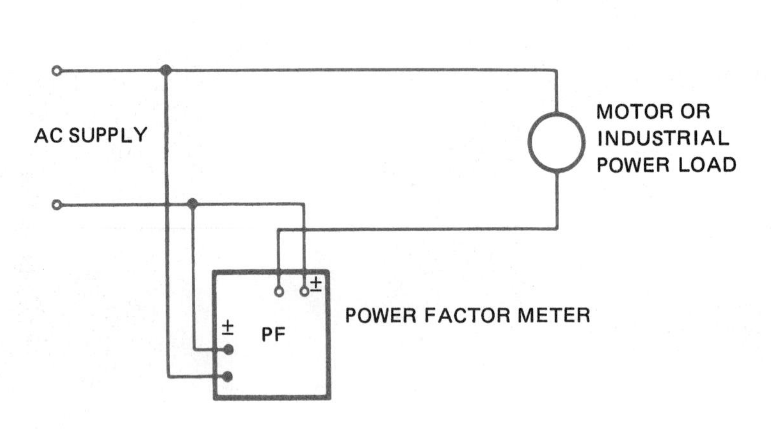

Fig. 9-18 Measurement of power factor using a power factor meter.

Using a Voltmeter, Ammeter, and Wattmeter

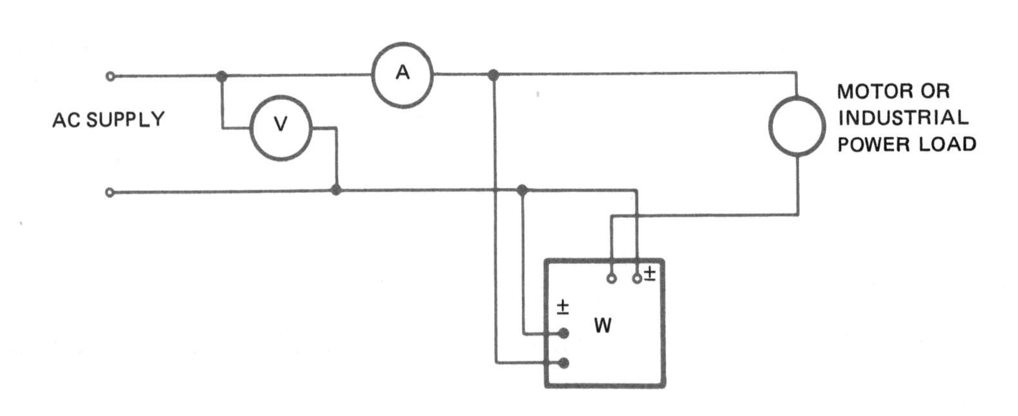

Fig. 9-19 Measurement of power factor using a voltmeter, ammeter, and wattmeter.

AC ELECTRICAL ENERGY

The watt (W) is the unit for power, or the rate at which energy is used. This unit is used for both direct- and alternating-current circuits.

The amount of energy used by a motor or the amount supplied to a branch line is measured in kilowatt-hours (kWh). Therefore, energy = power × time, and energy, not power, is the quantity purchased from a utility company.

Fig. 9-20 A portable ac voltmeter. *(Courtesy Weston Instruments, Inc.)*

Fig. 9-21 Digital clamp-on multimeter (voltmeter-ammeter-ohmmeter). *(Courtesy Amprobe Instrument Division)*

ACHIEVEMENT REVIEW

In items 1-6, select the *best* answer, and place the letter of the answer in the space provided.

1. Normally, the power factor of an incandescent lighting circuit is __________
   a. 0.
   b. 0.5.
   c. 0.707.
   d. 0.866.
   e. 1.0.

2. A 110-volt transformer draws 5 amperes and takes 440 watts. The apparent power, in volt-amperes, is __________
   a. 110.
   b. 115.
   c. 440.
   d. 550.
   e. 2 200.

3. Power companies are interested in improving the power factor to __________
   a. reduce line current.
   b. increase motor efficiency.
   c. reduce line voltage.
   d. increase volt-amperes.
   e. decrease power.

4. A capacitor increases the power factor value of an ac motor load when it is connected __________
   a. in series with the motor.
   b. across the starter winding.
   c. in parallel with the motor.
   d. in series with the main winding.
   e. across the series resistance.

5. In an ac parallel RL circuit, the power is developed at the __________
   a. impedance.
   b. resistance.
   c. inductance.
   d. capacitance.
   e. load.

6. In an ac series resistive circuit, for each cycle of source voltage the power waveform completes __________
   a. 1/4 cycle.
   b. 1/2 cycle.
   c. 1 cycle.
   d. 2 cycles.
   e. 4 cycles.

Problems 7–10 refer to the following diagram.

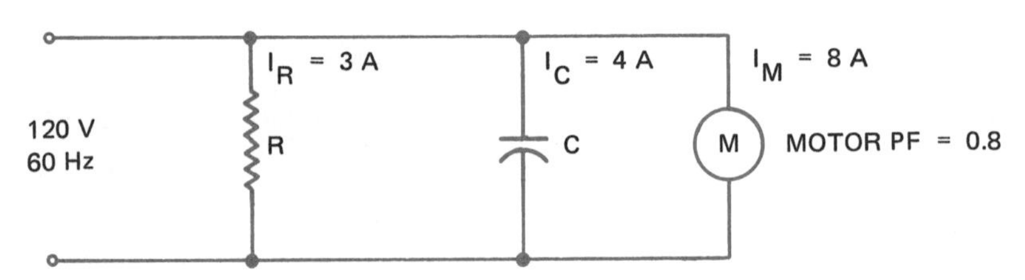

7. Which branch has a PF of 1, and which branch has a PF of 0?

8. Calculate the power and apparent power at resistor R.

9. Calculate the power and apparent power at the capacitor.

10. Calculate the power and apparent power at the motor.

11. A motor takes 49.2 W of power and draws 0.82 A from a 120-V, 60-Hz source. Find the power factor of the motor.

12. A welding transformer draws 80 A at 250 V. A wattmeter indicates that 16 kW of power is being taken by the transformer. It is desired to raise the power factor to 1.0 or 100 percent.

    a. Find the present power factor.

    b. What size capacitor, in kVA, is necessary to raise the power factor to 100 percent?

10 Summary Review of Units 1-9

OBJECTIVE

- To evaluate the knowledge and understanding acquired in the study of the previous nine units.

In items 1–10, insert the word or phrase that will make each incomplete statement true.

1. The unit that is used for inductance is the ____________________________ .

2. The value of current that is indicated by an ac ammeter is called the ____________ value.

3. Sixty hertz indicates sixty electrical cycles per __________________________ .

4. Sixty hertz is a measurement of ____________________________________ .

5. Three mechanical revolutions of a two-pole alternator result in ________________ electrical cycles.

6. The expression of $2\pi fL$ is used to determine ____________________________ .

7. The only component in a series RLC circuit that takes power is the ____________ ____.

8. In a series RLC circuit, when X_L is greater than X_c, the total current __________ ______________________ the total voltage.

9. When a capacitor is placed across an inductive load, the circuit power factor approaches the value of ____________________ .

10. With the condition of series resonance, total current and voltage are ____________ _________________________ .

In items 11–20, select the *best* answer to complete each statement. Place the letter of the selected answer in the space provided.

11. Inductive reactance is measured in ______________ .
12. Capacitance is measured in ______________ .
13. Impedance is measured in ______________ .
14. Power is measured in ______________ .
15. Frequency is measured in ______________ .
16. Capacitive reactance is measured in ______________ .
17. At resonance, the angle in degrees between total current and voltage is ______________ .
18. The power factor of a purely capacitive circuit is ______________ .
19. In a series resonant circuit, total current may be found with the expression ______________ .
20. The units for apparent power are ______________ .

a. henrys
b. seconds/cycle
c. volt-amperes
d. 0
e. farads
f. 1
g. 90°
h. $\frac{E_R}{R}$
i. 45°
j. ohms
k. kW
l. hertz
m. $I_T(Z)$
n. $I_T(R)$
o. 0.5

In items 21–45, select the *best* answer to make each incomplete statement true. Place the letter of the selected answer in the space provided.

21. In an incandescent lighting circuit, the total current generally ______________
    a. lags the voltage by 90°.
    b. leads the voltage by 90°.
    c. lags the voltage by 45°.
    d. is in phase with the voltage.

22. In an RL circuit, the total current ______________
    a. lags the total voltage by less than 90°.
    b. lags the total voltage by 45°.
    c. is in phase with the total voltage.
    d. lags the total voltage by 90°.

23. If line current in an ac series circuit leads line voltage by 60°, ______________
    a. $R = X_L$.
    b. X_L is greater than X_C.
    c. X_C is greater than X_L.
    d. $X_L = X_C$.

24. The impedance of an RL circuit may be found by using ______________
    a. $R + X_L$.
    b. (E)(R).
    c. $R^2 + X_L$.
    d. $\sqrt{R^2 + X_L^2}$.

25. The phase relationship between voltage and current at a capacitor is that the current ________
    a. leads the voltage by 45°.
    b. is in phase with the voltage.
    c. leads the voltage by 90°.
    d. lags the voltage by 90°.

26. In a resonant series circuit, the ________
    a. line current has a maximum value.
    b. total reactance has a maximum value.
    c. line current has a minimum value.
    d. resistance equals the reactance.

27. Inductive reactance is directly related to ________
    a. resistance.
    b. frequency.
    c. capacitance.
    d. power.

28. The expression for capacitive reactance in an RC circuit is ________
    a. $2\pi fL$.
    b. $X_L - X_C$.
    c. $\frac{1}{2\pi fC}$.
    d. E_T/I_T.

29. If the angle between total current and voltage is zero degrees, the ________
    a. current leads the voltage.
    b. current lags the voltage.
    c. voltage has a higher value than the current.
    d. current and the voltage are in phase.

30. When resistance alone is used to determine current in an RLC series circuit, the circuit is ________
    a. an inductive circuit.
    b. a capacitive circuit.
    c. a combination circuit.
    d. a resonant circuit.

31. The principal advantage of power factor correction is that the ________
    a. capacitor current is in phase with the source voltage.
    b. line current has a minimum value.
    c. load voltage increases.
    d. motors run faster.

32. A parallel circuit is in resonance when the ________
    a. inductive and capacitive branch currents are equal.
    b. line current has a maximum value.
    c. current in the resistive branch has a maximum value.
    d. line current and voltage are slightly out of phase.

33. For a circuit in which the total current lags the total voltage, the total circuit power factor can be raised to unity (or 100 percent) by __________
    a. connecting a resistor in series with the source.
    b. connecting an inductor in series with the source.
    c. connecting a capacitor across the line.
    d. increasing the inductance.

34. The power in an ac circuit may be calculated by using __________
    a. $E_T \times I_T$.
    b. AP $\times$ PF.
    c. $I_T \times R$.
    d. $\frac{E_T}{I_T}$.

35. Power factor is __________
    a. $\frac{R}{X}$.
    b. $\frac{\text{Power}}{\text{Apparent Power}}$.
    c. the efficiency.
    d. the ratio of reactive current to in-phase current.

36. A power factor of 1.0 exists only when the __________
    a. current leads the voltage.
    b. current lags the voltage.
    c. inductance equals the resistance.
    d. voltage and the current are in phase.

37. In an incandescent lighting circuit, __________
    a. apparent power equals power.
    b. resistance is equal to reactance.
    c. current lags voltage.
    d. the power factor is zero.

38. A transformer with a power factor of 0.9 takes 2 amperes at 100 volts. The power, in watts, is __________
    a. 50. c. 180.
    b. 90. d. 200.

39. The motor line meters indicate that the current is 8 amperes, the voltage is 120 volts, and the power is 768 watts. The power factor is __________
    a. 1.0. c. 0.64.
    b. 0.8. d. 0.156.

40. A series circuit includes R = 3 ohms, X_L = 4 ohms, and X_c = 4 ohms. The total impedance is __________
    a. 3 ohms. c. 11 ohms.
    b. 4 ohms. d. $\sqrt{3^2 + 8^2}$ ohms.

41. A parallel circuit has the following branch currents: I_R = 3 amperes, I_L = 6 amperes, I_C = 2 amperes. The value and phase of the line current is ______
a. 3 amperes in phase.
b. 5 amperes lagging.
c. 6 amperes lagging.
d. 11 amperes leading.

For problems 42–45, refer to the following diagram.

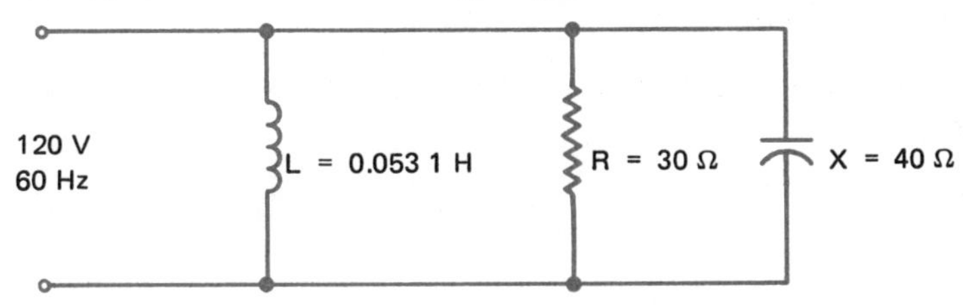

42. The current through the coil is ______
a. 6 amperes.
b. 20 amperes.
c. 22.6 amperes.
d. impossible to determine.

43. The current through the resistor is ______
a. 1.72 amperes.
b. 4.0 amperes.
c. 30 amperes.
d. 40 amperes.

44. The current through the capacitor is ______
a. 40 amperes.
b. 30 amperes.
c. 3.0 amperes.
d. 1.72 amperes.

45. The circuit is ______
a. a resonant circuit.
b. primarily a capacitive circuit.
c. primarily a resistive circuit.
d. primarily an inductive circuit.

11 Single-Phase, Three-Wire Service Entrance

OBJECTIVES

After studying this unit, the student will be able to

- analyze the requirements of a single-family dwelling.
- determine the size of service-entrance conductors.
- select the proper material and demonstrate the proper methods for the service-entrance installation.

The service entrance for most present-day lighting installations is a single-phase, three-wire service, figure 11-1.

The middle wire is called the neutral wire and is grounded. Therefore, this neutral wire (grounded conductor) is the identified or white wire of any single-phase, three-wire installation. The two outside wires are known as the hot wires (conductors). The voltage between the neutral wire and either of the two hot wires is 115 volts, and the voltage between the two hot wires is 230 volts.

It is an advantage to have both 115 and 230 volts available. Many types of loads, such as electric water heaters, electric ranges and fractional horsepower motors, operate on 230 volts.

Service-entrance voltages have gradually increased in many parts of the country. Therefore, it is not uncommon to find 120 volts and 240 volts as standard voltages.

TYPICAL SINGLE-FAMILY DWELLING

It is the intent of this unit to present the fundamental installation rules for a service entrance insofar as they concern the calculations that affect the service-entrance switch, service-entrance conductors, and grounding. The branch circuits supplying the various items of electrical equipment in a residence are covered briefly. The metering facilities for electrical space heating equipment as well as electric water heaters vary

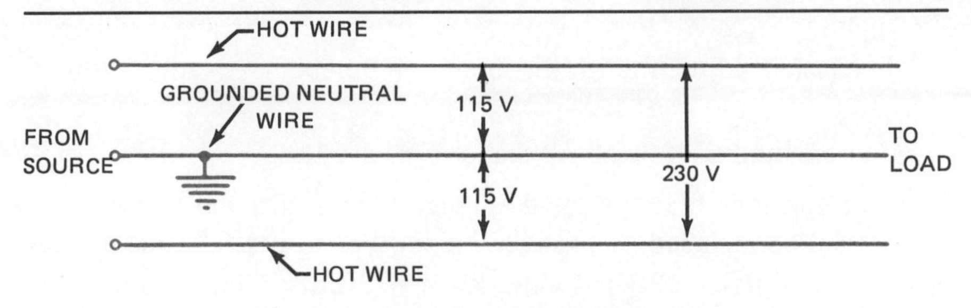

Fig. 11-1 Single-phase, three-wire system.

according to the requirements of the local utility company. As a result, it is not possible to cover all of these metering methods in detail. The electrician must check the requirements for these installations with the utility company serving the area in which the wiring is to be installed.

A typical application for a 115/230-volt, single-phase, three-wire service installation is a single-family dwelling. The residence considered in this unit is a six-room house (including three bedrooms) with an area of 1 500 square feet (1 500 ft^2). The residence contains a 3-kW water heater; a 5.0-kW clothes dryer; an 8-A, 230-V room air conditioner; a dishwasher rated at 11.1 A, 115 V; a 4-kW wallmounted oven; a 6-kW countermounted cooking unit; a garbage disposal rated at 7.5 A, 115 V; and 12 kW of electric space heating equipment installed for all six rooms. Each room has individual thermostatic control.

Determination of Number of Lighting Circuits

To determine the number of lighting circuits, the lighting load of the residence must be calculated. The calculations are based on the watts per square foot method outlined in the National Electrical Code (referred to here as "the Code"). In general, the outside dimensions of the building are used, not including open porches, garages, unfinished attics, or basements. For the residence in question, the area is 1 500 square feet. The recommended unit load is 3 watts per square foot.

Therefore, the total lighting load in watts is:

$$1\,500 \text{ square feet} \times 3 \text{ watts per square foot} = 4\,500 \text{ watts}$$

To determine the minimum number of 115-volt branch lighting circuits:

$$\text{amperes} = \frac{\text{watts}}{\text{volts}} = \frac{4\,500}{115} = 39.1 \text{ or } 40 \text{ amperes}$$

In general, 15-ampere lighting circuits using No. 14 AWG conductors are installed in residential occupancies. Some electrical specifications require a minimum conductor size of No. 12 AWG on all circuits.

$$\text{Thus: } \frac{40}{15} = 2 \text{ plus or 3 lighting circuits (minimum)}$$

However, a residence of this type may have as many as 60 outlets, including ceiling fixtures, porch fixtures, and wall convenience receptacles located throughout the living area, basement, garage, and grounds. As a result, most electricians prefer to limit the number of outlets per circuit to 8 or 10. This results in a more adequate number of lighting circuits. For the residence covered in this unit, at least 6 lighting circuits will be installed even though the minimum number of circuits required is 3.

Determination of Number of Small Appliance Circuits

The Code specifies that for small appliances an additional load of not less than 1 500 watts shall be included for each circuit for the receptacle outlets. These circuits shall feed only receptacle outlets in the kitchen, pantry, family room, dining room, and breakfast room of a dwelling. Two or more 20-ampere branch circuits shall be provided, and such circuits shall have no other outlets. No. 12 AWG wire is used instead of No. 14 AWG wire to minimize the voltage drop in the circuit. Thus, by using the larger wire, the performance of appliances is improved and the danger of overloading circuits is decreased.

Since automatic washing machines draw a large amount of current during certain portions of their operating cycles, the Code requires that a separate 20-ampere circuit for the laundry outlets be installed. All convenience receptacles must be of the grounding type.

To determine the service-entrance requirements of this dwelling, the small appliance load is assumed to be 4 500 watts, based on three 20-ampere circuits at 1 500 watts per circuit.

Garbage Disposal. The garbage disposal unit is rated at 7.5 amperes, 115 volts, and will be supplied by a separate 15-ampere circuit which requires No. 14 AWG conductors.

The garbage disposal load is:

$$7.5 \text{ amperes} \times 115 \text{ volts} = 862 \text{ watts}$$

Dishwasher. The dishwasher is rated at 11.1 amperes, 115 volts, and will be supplied by a separate 15-ampere circuit using No. 14 AWG wire. (It is possible to supply the garbage disposal unit and the dishwasher with one 115/230-volt, three-wire circuit.)

The dishwasher load in watts is:

$$11.1 \text{ amperes} \times 115 \text{ volts} = 1\ 276 \text{ watts}$$

Dryer Circuit. The electric clothes dryer is rated at 5.0 kW, 230 V. The current it draws is:

$$\frac{5\ 000 \text{ watts}}{230 \text{ volts}} = 21.7 \text{ amperes}$$

The Code states that the branch-circuit rating for continuous duty loads shall be not less than 125 percent of the rating of the appliance.

$$21.7 \text{ amperes} \times 1.25 = 27.1 \text{ amperes}$$

The circuit to the dryer will be a 30-ampere, 230-volt, three-wire circuit consisting of two No. 10 AWG conductors for the hot wires and one No. 10 AWG conductor for the neutral conductor.

Wallmounted Oven. The oven is rated at 4 kW, 115/230 V and will be connected to a separate circuit. In watts, the load is 4 000 watts; in amperes, the load is equal to:

$$\text{amperes} = \frac{\text{watts}}{\text{volts}} = \frac{4\ 000}{230} = 17.4 \text{ amperes}$$

No. 12 AWG wire may be used for this circuit.

Countermounted Cooking Unit. The surface cooking unit is rated at 6 kW, 115/230 V. The unit will be connected to a separate circuit consisting of three No. 10 AWG conductors supplied by a 30-ampere, two-pole overcurrent device.

The load, in amperes, is:

$$\text{amperes} = \frac{\text{watts}}{\text{volts}} = \frac{6\ 000}{230} = 26 \text{ amperes}$$

Air Conditioner. The room air conditioner draws 8 amperes at 230 volts and will be connected to a separate 15-ampere, 230-volt circuit with No. 14 AWG conductors (15 amperes).

The air conditioner load, in watts, is:

$$8 \text{ amperes} \times 230 \text{ volts} = 1\,840 \text{ watts}$$

Water Heater. Many utility companies furnish current for residential electric water heater loads at a power consumption rate lower than the regular lighting rate. In such installations, some utility companies require a separate *off-peak* meter, while other companies predetermine a fixed portion of the monthly light bill to cover the power consumption of the water heater.

In general, for the off-peak metering circuit, the top element of the heater is connected to a two-pole, 230-volt circuit supplied through the house meter. The bottom element of the heater is connected to a two-pole, 230-volt circuit supplied through the off-peak meter. These elements can be connected for *limited demand*, in which case both elements cannot be energized simultaneously; or they may be connected for *unlimited demand*, in which case both elements may be energized simultaneously. The types of thermostats furnished with the water heater determine how the elements are connected.

Various types of equipment are manufactured in which both regular and off-peak overcurrent protective devices may be located in the same enclosure. The off-peak device is called a *feedthrough unit* and is not connected in any manner to the main bus of the panel even though it is located in the same enclosure. In the feedthrough unit, the two wires from the off-peak meter are connected to one side of the unit, and the two wires supplying the element of the water heater are connected to the other side.

The water heater in this residence is rated at 3 kW. This load, in amperes, is equal to:

$$\text{amperes} = \frac{\text{watts}}{\text{volts}} = \frac{3\,000}{230} = 13 \text{ amperes}$$

When connected for unlimited demand, the maximum current of this water heater is 13 amperes.

Consult the local utility company for guidelines on the proper connection of water heaters.

Electric Space Heating. The specified total of 12 kW of electric space heating units will be installed throughout the residence. Each of the six rooms will have a thermostat to provide individual heating control. According to the National Electrical Code, there must be four or more individually controlled electric space heating units to apply certain demand factors permitted by the Code. These factors are used to calculate the service-entrance capacity. Approximately 2 kW of space heat will be provided in each room. Since these units are rated at 230 volts, the load for each is:

$$\text{amperes} = \frac{\text{watts}}{\text{volts}} = \frac{2\,000}{230} = 8.7 \text{ amperes}$$

The branch-circuit current rating is 125 percent of the load, or, 1.25 × 8.7 = 10.9 amperes. Therefore, each unit will be connected to a separate 15-ampere, 230-volt circuit using No. 14 AWG wire.

Some utility companies offer lower rates for electric heating when this is a residential requirement in addition to general electrical services. These rates usually are based on special metering methods. The electrician should consult the utility company supplying power in a local area for the correct method of connecting heating loads.

Summary of Branch Circuits for the Residence					
No. of Circuits	Voltage	Use	Branch-Circuit Ampere Rating	Poles	Wire Size (AWG No.)
6	115	General lighting	15	1	14
3	115	Small appliance & laundry	20	1	12
1	115	Garbage disposal	15	1	14
1	115	Dishwasher	15	1	14
1	115/230	Dryer	30	2	10
1	115/230	Oven	20	2	12
1	115/230	Surface cooking unit	30	2	10
1	230	Air conditioner	15	2	14
1	230	Water heater	20	2	12
6	230	Space heat	15	2	14

SIZE OF SERVICE-ENTRANCE CONDUCTORS

Section 230-41 of the National Electrical Code specifies that service-entrance conductors shall have a current-carrying capacity sufficient to carry the load as determined by *Article 220*. For dwelling occupancies, the Code permits the use of either of two methods to determine the size of these conductors.

Method Number 1 (Standard Method) (*Article 220, Parts A and B*)

General Lighting Load:

1 500 square feet at 3 watts per square foot . . . . 4 500 watts

Small Appliance Load (*Section 220-16*):

Three 20-ampere appliance circuits at 1 500 watts per circuit . . . . 4 500 watts

Total (without range) 9 000 watts

Application of demand factor (*Table 220-11*):

3 000 watts at 100% = ... 3 000 watts

9 000 – 3 000 = 6 000 watts at 35% = ... 2 100 watts

Net computed load without range 5 100 watts

Wallmounted Oven and Countermounted Cooking Unit
(*Table 220-19, Note 4*):

6 000 + 4 000 = 10 000 watts at 80% = ...	8 000 watts
Net computed load with range (5 100 + 8 000) =	13 100 watts
Electric Space Heating (*Section 220-15*)	12 000 watts

Air conditioner wattage is 8 X 230 = 1 840 watts. This value is less than the 12 000 watts of space heating; therefore, the air conditioner load need not be included in the service calculation (*Section 220-21*).

Water Heater	3 000 watts
Dryer	5 000 watts
Dishwasher	1 276 watts
Garbage Disposal (862 X 1.25) [*Section 210-22(a)*]	1 077 watts
Total	10 353 watts

Since there are four appliances in addition to the cooking units and space heating, a demand factor of 75 percent may be applied to the fixed appliance load (*Section 220-17*).

Thus, 10 353 X 0.75 = ...	7 765 watts
Total Calculated Load: 13 100 + 12 000 + 7 765 = ...	32 865 watts

$$\text{Amperes} = \frac{\text{watts}}{\text{volts}} = \frac{32\,865}{230} = 142.9 \text{ amperes}$$

According to *Table 310-16 Note 3* of the Code, for a load of 142.9 amperes, No. 1 RHW or THW wire may be installed as the copper service-entrance conductors.

It should be noted that for single-family residences with initial load of 10 kW or more, computed in accordance with *Article 220*, there shall be a minimum of a 100-ampere, three-wire service. This minimum also applies to residences that have six or more two-wire branch circuits [*Section 230-4(b)*].

Method Number 2 (*Section 220-30*)

An optional method of determining the load of a single-family dwelling is recognized by the Code. This method simplifies the calculations and usually results in a smaller size of service entrance than is permitted by method 1.

1 500 square feet at 3 watts per square foot	4 500 watts
Three 20-ampere appliance circuits at 1 500 watts per circuit. . .	4 500 watts
Wallmounted oven (nameplate rating).	4 000 watts
Countermounted cooking unit (nameplate rating).	6 000 watts
Water heater	3 000 watts
Dryer.	5 000 watts
Dishwasher	1 276 watts
Garbage disposal (862 X 1.25) [*Section 210-22(a)*]	1 077 watts
Electric space heating	12 000 watts
Air conditioner wattage is 8 X 230 = 1 840 watts. This value is less than the 12 000 watts of space heating; therefore, the air conditioner load need not be included in the service calculation. [*Section 220-21*]	
Total load	41 353 watts

Then: first 10 kW at 100% =	10 000 watts
remainder of load at 40% (31 353 X 0.4) =	12 541 watts
Total Calculated Load	22 541 watts

$$\text{Amperes} = \frac{\text{watts}}{\text{volts}} = \frac{22\ 541}{230} = 98 \text{ amperes}$$

According to *Table 310-16* of the Code, for a load of 98 amperes, No. 4 RHW wire or equivalent may be installed as the copper service-entrance conductors.

Both Methods 1 and 2 for determining total load are correct as far as the Code is concerned. Therefore, the decision as to which method is permitted in an area is made by the local electrical inspector.

To provide a single panel which will accommodate all the circuits in the residence, it is necessary to install a 200-ampere panel.

Certain localities require that the conductors supplying a panel or switch must have a current-carrying capacity equal to the rating of the panel or switch. Therefore, for the residence covered in this unit, No. 2/0 RHW or equivalent wire is required for the service entrance. The installation of No. 2/0 RHW wire or equivalent will give the homeowner full 200-ampere capacity. See *Note 3* to *Table 310-16.*

Service-Entrance Switch (*Sections 230-70 and 230-71*)

Section 230-71 of the National Electrical Code in essence specifies that the service disconnecting means shall consist of not more than six switches or six circuit breakers in a single enclosure or in a group of separate enclosures. It is the intent of this section to insure that all electrical equipment within a building can be disconnected with no more than six manual operations. However, certain local ordinances do not permit the six subdivision rule but rather require that each service shall have a single main disconnect.

To accommodate the number of circuits listed in the Summary of Branch Circuits, page 69, a 200-ampere panel will be installed. This panel will contain all of the required branch circuits plus a 200-ampere main pullout in one enclosure. This type of enclosure is acceptable as both the load center and service equipment, and meets Code requirements in most localities.

Generally, the service switch is located in the basement and the meter is mounted on the outside of the house for easy access by the utility company.

Ground Connection

Section 250-5(b)(1) of the Code requires the grounding of interior alternating-current systems where the system can be grounded so that the maximum voltage to ground, on the ungrounded conductors, does not exceed 150 volts. Grounding is accomplished by running a wire from the neutral connection in the main service switch or meter to the water piping system on the street side of the water meter. The reason for connecting on the street side of the water meter is so that the ground circuit remains connected if the meter must be removed for repair.

Sections 250-91 and *250-92* set forth the rules governing grounding materials and the installation of the ground wire. The size of ground wire required is found in *Table 250-94*. It was mentioned previously that the residence covered in this unit will be supplied with No. 2/0 RHW service-entrance conductors. According to *Table 250-94*, No. 2/0 RHW conductors require a No. 4 AWG grounding conductor.

Note: It is beyond the scope of this unit to illustrate all the methods of metering the water heater and electrical space heating load. Figure 11-2, therefore, illustrates the entire load connected to a single meter. This figure is used only to outline the installation requirements.

Bonding

The proper bonding of all service-entrance equipment is as important as the use of the proper size of service conductors. *Section 250-71* lists the equipment that shall be bonded and *Section 250-72* lists the methods approved for bonding this equipment. Bonding jumpers shall have a current-carrying capacity not less than is required for the corresponding grounding conductor.

The purpose of bonding on service-entrance equipment is to assure a low impedance path to ground should a fault occur on any of the service-entrance conductors. **Severe arcing, which presents a fire hazard, may occur at a fault.** Proper bonding reduces this hazard to some extent.

The fire hazard exists because the service-entrance conductors are not fused at the service head. The short-circuit current on these conductors is limited only by the capacity of the transformer or transformers supplying the service equipment and the distance the service equipment is located from these transformers. Short-circuit current can easily reach 10 000 amperes in residential areas and as high as 200 000 amperes in industrial areas. All overcurrent devices (fuses and circuit breakers) must have adequate interrupting capacity. See *Sections 110-9* and *230-98* of the Code.

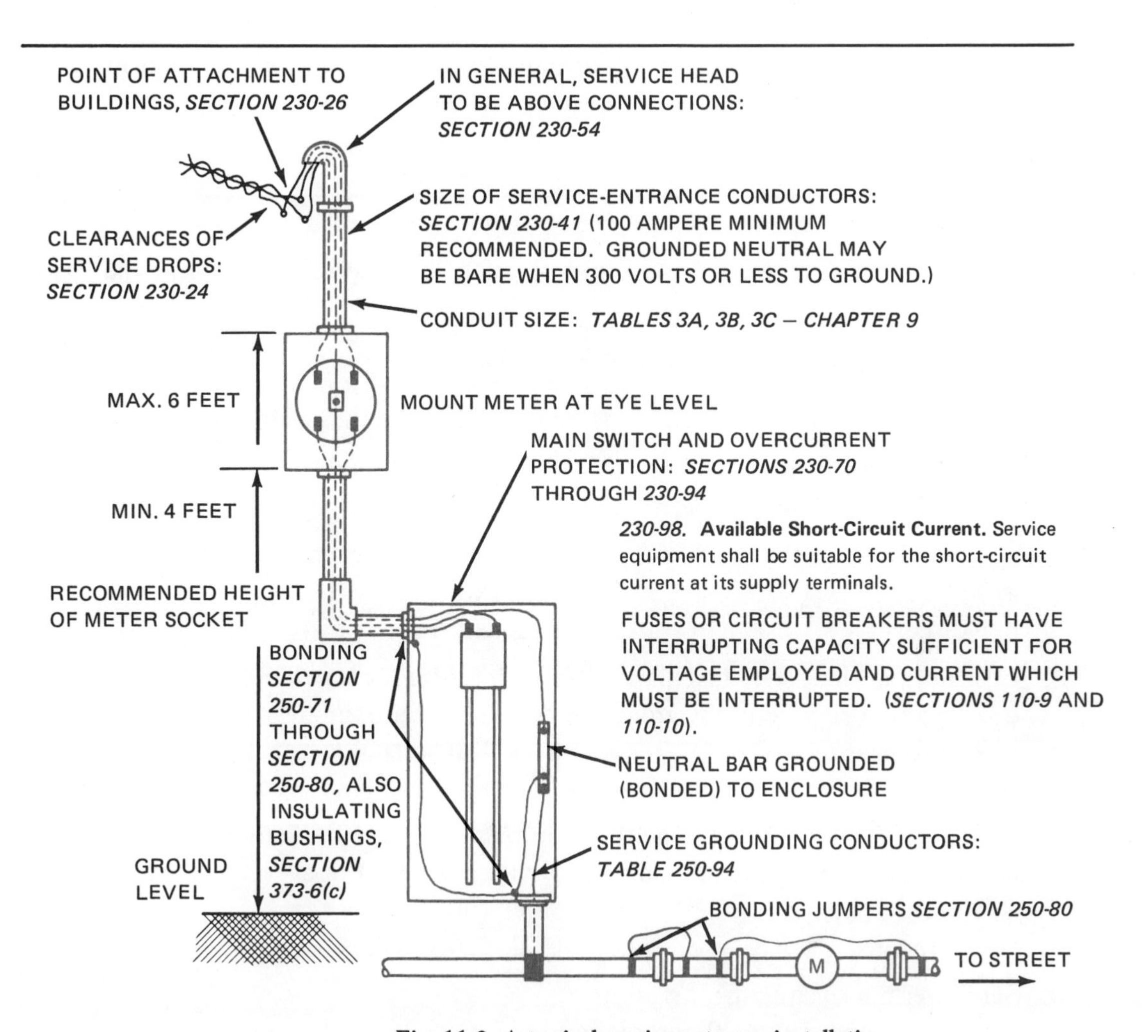

Fig. 11-2 A typical service-entrance installation.

ACHIEVEMENT REVIEW

1. State one reason why the single-phase, 115/230-volt, three-wire system is used in present-day installations instead of a 115-volt, two-wire system.

 ____

 ____

 ____

2. State how the area, in square feet, for a single-family dwelling is determined to arrive at the approximate lighting load.

 ____

 ____

3. What is the number of watts per square foot allowed by the Code when determining the lighting load for a single-family dwelling? ______________

4. What is the minimum number of appliance circuits permitted by the Code for a single-family dwelling? ______________

5. Explain why the ground wire is connected to the street side of the water meter in installations where there is a public water system.

____

____

In items 6–10, select the *best* answer, and place the corresponding letter in the space provided.

6. A single-phase, 115/230-volt, three-wire system shall ________
   a. be grounded.
   b. be ungrounded.
   c. have one appliance circuit.
   d. have No. 12 service-entrance wires.

7. The wire used for the small appliance circuits cannot be smaller than ________
   a. No. 10.
   b. No. 12.
   c. No. 14.
   d. No. 16.

8. In a normal three-wire installation for a single-family dwelling, the voltage from one ungrounded wire to the grounded neutral is approximately ________
   a. 460 volts.
   b. 230 volts.
   c. 150 volts.
   d. 115 volts.

9. For single-family residences with an initial load of 10 kW or more, computed according to acceptable methods, the service shall be at least ________
   a. 50 amperes.
   b. 80 amperes.
   c. 100 amperes.
   d. 120 amperes.

10. Service-entrance equipment shall be ________
    a. of a common size for all installations.
    b. bonded.
    c. stapled.
    d. selected before the load is determined.

11. Determine the minimum number of 115-volt lighting circuits necessary in a single-family dwelling with an active area of 2 300 square feet.

____

____

____

____

12. A single-family dwelling has a 7.6-kW, 115/230-volt electric range. Determine the minimum size of the conductors to be used for the range feeder from the service-entrance cabinet to the range outlet.

____

____

____

12 Installation of a Single-Phase, Three-Wire Service Entrance for an Apartment Building

OBJECTIVES

After studying this unit, the student will be able to

- explain the connections for a single-phase, three-wire entrance for an apartment building.
- compute the size of the subfeeders to individual apartments.
- compute the size of the service wires.

The construction of new apartment buildings and the conversion of older buildings into apartment dwellings is a continuing trend in urban areas. It is often the job of the electrician to install the single-phase, three-wire service entrance for an apartment building. The electrician must determine the size of the conductors and conduit for the service entrance and the size of the subfeeders for each apartment.

In this unit, the problem is to install the service entrance for a building containing twenty apartments. Certain sections of this unit are similar to those of unit 11.

APARTMENT BUILDING INSTALLATION

Assume that each of the twenty apartments in this building has a floor area of 800 square feet. Ten of these apartments have 9-kilowatt ranges.

The main single-phase, three-wire service will enter the building in the basement. The main service-entrance switch and the individual meters for each of the twenty apartments will be located in the basement. There will be individual feeders from each meter to its respective apartment. A panel located in each apartment will serve as the distribution point for the individual branch lighting circuits, small appliance circuits, and the range circuit.

Computation of the Load in Each Apartment without a Range

Branch Lighting Circuits. The Code specifies that the active area in square feet shall be computed from the outside dimensions of the apartment, not including open porches, garages, or unfinished spaces in attics. The floor area of each apartment is known to be 800 square feet.

The general lighting load is determined by the watts per square foot method. The total lighting load for each apartment is:

800 square feet × 3 watts per square foot = 2 400 watts

To determine the number of 115-volt, two-wire branch lighting circuits:

$$\text{amperes} = \frac{\text{watts}}{\text{volts}} = \frac{2\,400}{115} = 20.9 \text{ or } 21 \text{ amperes}$$

Then, the minimum number of 15-ampere, 115-volt circuits is,

$$\frac{21}{15} = 1 \text{ plus or } 2 \text{ lighting circuits}$$

Small Appliance Circuits. The Code specifies that in addition to the general lighting circuits, two or more 20-ampere branch circuits shall be provided for the outlets in the kitchen, pantry, family room, dining room, and breakfast room; and further, that such circuits shall supply no other outlets. The required wire size for these circuits is No. 12 AWG.

For simplicity in the following calculations, it is assumed that there are no provisions for laundry facilities in the apartment building. Therefore, an additional load will not be added for laundry circuits. (If laundry facilities were to be provided, an additional load of 1 500 watts per circuit would be required.)

Subfeeder. The Code specifies that the load used to determine the size of subfeeder conductors shall be computed as follows:

General Lighting Load.		2 400 watts
Small Appliance Load		3 000 watts
	Total	5 400 watts
Application of Demand Factor: (*Table 220-11*)		
3 000 watts.	at 100%	3 000 watts
5 400 − 3 000 = 2 400 watts.	at 35%	840 watts
	Net computed load	3 840 watts

$$\text{Thus, amperes} = \frac{\text{watts}}{\text{volts}} = \frac{3\,840}{230} = 16.7 \text{ amperes}$$

Code *Section 215-2* states that for a three-wire feeder supplying more than two two-wire branch circuits, or two or more three-wire branch circuits, the feeder shall not be smaller than No. 10 AWG.

Therefore, the subfeeder to each of the ten apartments which do not contain an electric range will consist of three No. 10 AWG copper conductors. Most electrical ordinances require conduit for an installation such as this. *Table 3A* in *Chapter 9* of the Code shows that three No. 10 TW conductors require 1/2-inch conduit.

Summary. In the ten apartments which do not have electric ranges, each will have:

- One 115/230-volt, three-wire, single-phase subfeeder using No. 10 AWG wire which will feed from the apartment's disconnect in the basement to the load center in the individual apartment.
- Two 115-volt, two-wire branch circuits for the lighting circuits. Generally, these circuits are 15-ampere circuits using either No. 14 or No. 12 wire.
- Two 115-volt, 20-ampere, two-wire circuits using No. 12 wire for the small appliance load.

Computation of the Load in Each Apartment with a Range

Each of the ten apartments with electric ranges also has an active floor area of 800 square feet. The connected lighting and appliance load is the same for these apartments as for the ten apartments without electric ranges. The rating of each range in these apartments is 9 kW.

Table 220-19 of the Code indicates that for a single, household, electric range rated at not over 12 kW, the maximum demand may be based on 8 kW (See *Column A* of *Table 220-19*). This load in amperes is:

$$\text{amperes} = \frac{\text{watts}}{\text{volts}} = \frac{8\,000}{230} = 34.8 \text{ amperes}$$

Therefore, the two hot wires to the range outlet will be No. 8 TW. According to *Section 210-19(b), Exception 1,* the neutral conductor may be smaller than the ungrounded conductors but shall have not less than 70 percent of the current-carrying capacity of the ungrounded conductors. No. 8 TW wire is rated at 40 amperes.

$$40 \text{ amperes} \times 0.70 = 28 \text{ amperes}$$

Therefore, the neutral conductor may be No. 10 TW wire (See *Table 310-16*).

When installing the range circuit in conduit, it is necessary to refer to the *Tables* in *Chapter 9* in the National Electrical Code. When conductors of different sizes are installed in a conduit, the cross-sectional area of the conductors shall not exceed the allowable percentage of fill of the interior cross-sectional area of the conduit as shown in the *Tables.*

From *Table 5*:

Two No. 8 TW conductors		0.047 1 in²
		0.047 1 in²
One No. 10 TW conductor		0.022 4 in²
	Total	0.116 6 in²

Checking *Table 4*, it is found that a 1/2-inch conduit will hold up to 0.12 in² conductor fill and a 3/4-inch conduit will hold up to 0.21 in² conductor fill. In this installation, 3/4-inch conduit will be installed for ease in pulling wires; however, the Code permits 1/2-inch conduit to be installed.

Subfeeder. *Section 220-10* of the Code states that the load used to determine the size of subfeeder conductors shall be computed as follows:

General Lighting Load.		2 400 watts
Small Appliance Load		3 000 watts
Total		5 400 watts
Application of Demand Factor: (*Table 220-11*)		
3 000 watts.	at 100%	3 000 watts
5 400 - 3 000 = 2 400 watts	at 35%	840 watts
Net computed load		3 840 watts
Range Load (*Table 220-19*).		8 000 watts
Total computed load (with range)		11 840 watts

$$\text{Thus, amperes} = \frac{\text{watts}}{\text{volts}} = \frac{11\,840}{230} = 51.5 \text{ amperes}$$

Checking *Table 310-16*, it can be seen that No. 6 TW wire is required for the ungrounded conductors.

Subfeeder Neutral. The size of the neutral conductor is computed as follows:

General lighting and small appliance load after applying demand factor (shown above).	3 840 watts
Range Load: 8 000 watts at 70% (See *Section 220-22*)	5 600 watts
Total computed neutral load	9 440 watts

$$\text{Thus, amperes} = \frac{\text{watts}}{\text{volts}} = \frac{9\,440}{230} = 41 \text{ amperes}$$

Checking *Table 310-16* shows that No. 6 TW wire will be required for the neutral conductor of the feeder.

To summarize, therefore, three No. 6 TW conductors will be installed for each apartment. According to *Table 3A* in *Chapter 9* of the Code, three No. 6 TW conductors require a 1-inch conduit.

Summary. For each of the ten apartments which do have electric ranges, there will be:

- One 115/230-volt, three-wire, single-phase subfeeder using No. 6 TW wire which will feed from the apartment's disconnect in the basement to the load center in the individual apartment.
- Two 115-volt, two-wire branch circuits for the lighting circuits. Generally, these circuits are 15-ampere circuits using either No. 14 AWG or No. 12 AWG wire.
- Two 115-volt, 20-ampere, two-wire circuits using No. 12 AWG wire for the small appliance load.
- One 115/230-volt, three-wire, single-phase circuit consisting of two No. 8 TW ungrounded conductors and one No. 10 TW neutral conductor for the range.

MAIN SERVICE-ENTRANCE CONDUCTORS

The main service-entrance conductors are computed as follows:

General Lighting and Small Appliance Load: 20 apartments X 5 400 watts equals		108 000 watts
Application of Demand Factor: (*Table 220-11*)		
3 000 watts.	at 100%	3 000 watts
108 000 - 3 000 = 105 000 watts	at 35%	36 750 watts
	Net computed load	39 750 watts
Range Load (ten 9-kW ranges): (See *Table 220-19, Column A.*)		25 000 watts
Total computed load (with ranges)		64 750 watts

$$\text{Thus, amperes} = \frac{\text{watts}}{\text{volts}} = \frac{64\,750}{230} = 281.5 \text{ amperes}$$

Checking *Table 310-16* shows that No. 300 MCM-RHW wire or equivalent can be installed for the ungrounded conductors.

SERVICE-ENTRANCE NEUTRAL

The service-entrance neutral is computed as follows:

General Lighting and Small Appliance Load after
applying demand factor: . . . . 39 750 watts
Range Load: 25 000 watts at 70% (*Section 220-22*) . . . . 17 500 watts
Total computed neutral load. . . . . 57 250 watts

Then, amperes $= \frac{\text{watts}}{\text{volts}} = \frac{57\,250}{230} = 249$ amperes

Further Demand Factor (*Section 220-22*)
200 amperes . . . . at 100% 200.0 amperes
249 - 200 = 49 amperes . . . . at 70% 34.3 amperes
Final computed neutral load . . . . 234.3 amperes

Checking *Table 310-16*, it is evident that No. 250 MCM-RHW wire or equivalent can be installed for the neutral conductor. The neutral conductor is permitted to be bare.

To determine the proper size conduit: (See *Table 5, Chapter 9* for cross-sectional areas)

		Cross-sectional area
Two No. 300 MCM-RHW conductors		0.683 7 in^2
		0.683 7 in^2
One No. 250 MCM-RHW conductor		0.591 7 in^2
	Total	1.959 1 in^2

Checking *Table 4, Chapter 9,* the proper size conduit for the service-entrance conductors is given as 3-inch conduit.

If a bare neutral is installed:

Two No. 300 MCM-RHW conductors		0.683 7 in^2	
		0.683 7 in^2	
One No. 250 MCM bare conductor		0.260 0 in^2	(*Table 8*)
	Total	1.627 4 in^2	

Therefore, 2 1/2-inch conduit may be used according to *Table 4, Chapter 9.*

SERVICE-ENTRANCE SWITCH

The requirements for the main service-entrance disconnecting means are contained in *Sections 230-70* through *230-76* of the Code. These sections require that a readily accessible means be provided to disconnect all ungrounded conductors in the building from the service-entrance conductors. This disconnecting means shall be externally operable; shall indicate clearly whether it is in the open or closed position; shall disconnect all ungrounded conductors simultaneously; and shall have a rating of not less than the load to be carried in accordance with *Article 220.* Generally, the overcurrent device shall not exceed the allowable current-carrying capacity of the service-entrance conductors (*Section 230-90*).

The main disconnect for this apartment will be a 400-ampere, three-wire solid neutral, 250-volt switch fused at 300 amperes. See *Sections 110-9* and *230-98* for interrupting capacity requirements.

GROUNDING OF SERVICE EQUIPMENT

Section 230-63 states that service equipment shall be grounded in the manner specified in *Article 250* (grounding). A common grounding conductor may be used for grounding the wiring system and the service equipment as provided in *Section 250-94.* For service conductors over 3/0 AWG to 350 000 circular mils inclusive, the required size of the grounding conductor is No. 2 AWG. If the grounding conductor is to be installed in conduit, it may be installed in 1/2-inch conduit (*Table 3A, Chapter 9*). Bonding requirements are found in *Sections 250-71* and *250-72*.

Figure 12-1 illustrates a typical service-entrance arrangement for an apartment building. Note that the service-entrance conduit feeds into the main entrance switch where the 300-ampere fuses are located. The switch is rated at 400 amperes.

Figure 12-1 also shows a short conduit nipple run from the main switch to an auxiliary gutter. The taps to each meter feed from the main conductors in the gutter. These taps must be large enough to carry the load required for each subfeeder. For each of the apartments without electric ranges, three No. 10 AWG wires will be used. For the taps feeding each of the apartments with electric ranges, three No. 6 TW wires will be used.

Overload protection for the subfeeders to each apartment is located in the disconnect switch above each meter. Each subfeeder is then run to the respective load centers located in each of the twenty apartments. Branch-circuit protection in the load centers may be in the form of breakers or fuses. If circuit breakers are used, they must conform to *Sections 240-80* through *240-83*. If fuses are used, they must conform to *Sections 240-50* through *240-61*.

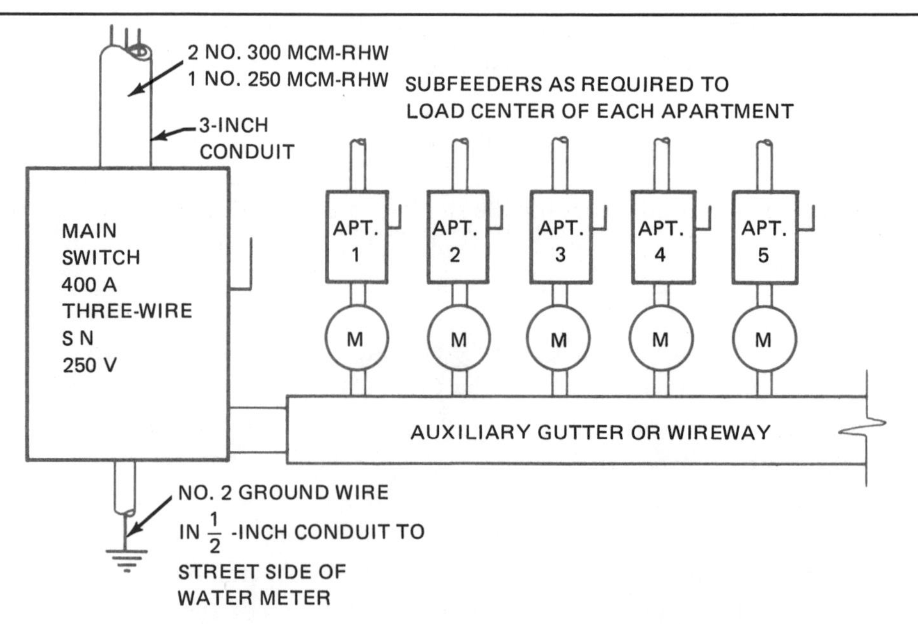

Fig. 12-1 Service entrance for apartment building.

ACHIEVEMENT REVIEW

1. What is the minimum number of watts per square foot permitted by the Code for apartment dwellings? ______________________

2. A certain service requires three No. 500 MCM-TW conductors. What is the minimum size conduit permitted for these conductors? ______________________

3. What size copper ground wire is necessary to ground the service in question 2?

   ____

4. What minimum load must be allowed for small appliances in apartment dwellings?

   ____

5. For a three-wire feeder supplying more than two two-wire branch circuits, what is the smallest size feeder wire that can be used? ______________________

6. A building has ten apartments. Each apartment has 600 square feet of active floor area. Five of these apartments will have 9-kilowatt, 115/230-volt electric ranges. Find the number of branch, 115-volt, two-wire lighting circuits required for each of the ten apartments. Also list the number of small appliance circuits. (Laundry facilities are not involved.)

7. Find the size of the single-phase, 115/230-volt, three-wire subfeeders to each of the five apartments without ranges given in question 6. Use TW wire.

8. Find the size of the single-phase, 115/230-volt, three-wire subfeeders to each of the five apartments with ranges given in question 6. Use TW wire.

9. Find the size of the conductors for the single-phase, 115/230-volt service entrance of the apartment building given in question 6. Use RHW wire.

10. If all three of the service-entrance conductors in question 9 are insulated (RHW), what is the minimum size conduit required?

13 Installation of a Three-Phase, Three-Wire Service Entrance

OBJECTIVES

After studying this unit, the student will be able to

- discuss the requirements for three-phase, three-wire service entrances not greater than 100 amperes.
- diagram the connections for a typical three-phase, three-wire service entrance.

Most power utility companies require a three-phase service installation when the connected motor load is in excess of 5 horsepower. Most three-phase motors with ratings of one horsepower or larger are less expensive than single-phase motors of equal ratings. Furthermore, a three-phase motor is easier to maintain than a single-phase motor. Single-phase motors usually have a centrifugal switch or, in some cases, a commutator, and thus require a great deal of maintenance. Three-phase motors have none of these devices to adjust or repair. The operating performance of three-phase motors in terms of torque, speed regulation, and efficiency is better than that of single-phase motors.

Industrial plants and other commercial facilities use a great many three-phase motors. Because there must be three-phase service entrances to each of these installations, it is important that the electrician know the requirements for three-phase service entrances.

PROCEDURE BEFORE STARTING WORK

Before starting work on three-phase installations, the electrician must give the power company detailed information as to the connected horsepower load.

The power company will then specify:

1. the location of the service entrance to the building and the location of the meter board.
2. the type and size of the meter test cabinet and test block.
3. installation standards.

INDUSTRIAL LOAD

The following example of a small industrial installation shows how to determine the size of a three-phase, 230-volt service entrance. The load consists of:

1. two branch circuits each feeding one squirrel-cage induction motor rated at 230 volts, 15 amperes, 5 horsepower, with no Code marking.
2. one branch circuit feeding a three-phase, squirrel-cage induction motor rated at 230 volts, 21 amperes, 7 1/2 horsepower, with no Code marking.

Size of Main Feeder

When conductors supply two or more motors, the Code gives definite instructions for determining the feeder size. *Section 430-24* specifies that the feeder shall have a current capacity of not less than 125 percent of the full-load current rating of the highest rated motor in the group, plus the sum of the full-load current ratings of the remainder of the motors in the group.

The largest motor in the example in this unit is rated at 7 1/2 hp with a full-load current rating of 21 amperes. To find the total current in compliance with the Code ruling:

125% of 21 = 1.25 X 21 = 26.25 amperes

Then, 26.25 + 15 + 15 = 56.25 amperes

Therefore, the main feeder can be a No. 4 TW wire which can carry 70 amperes, or a No. 6 RHW wire which can carry 65 amperes.

Size of Feeder Fuses (Short-circuit and Ground Fault Protection)

Section 430-62 (a) of the Code specifies that a feeder which supplies motors shall be provided with overcurrent protection. This overcurrent protection is determined by taking the current rating of the branch motor circuit with the largest fuse rating, plus the sum of the full-load currents to the other motors.

For this example, the branch circuit feeding the 7 1/2-hp motor has the largest fuse rating.

Table 430-152 of the Code gives the permissible factors to apply to the full-load ratings of different types of motors to determine the branch-circuit protection. The factor for three-phase, squirrel-cage induction motors without Code markings is 175 percent for time-delay fuses.

The branch circuit feeding the 7 1/2-hp motor will have the following fuse protection:

175% of 21 amperes = 1.75 X 21 = 36.75 amperes

Therefore, 40-ampere time-delay fuses and a 60-ampere switch can be used in this branch circuit.

It is possible to use time-delay fuses sized at 125 percent of the motor current rating.

1.25 X 21 = 26.25

Therefore, 25-ampere or 30-ampere time-delay fuses may be installed in a 30-ampere switch.

Section 430-32 specifies that the running overload protection shall be not more than 125 percent of the full-load current rating of the motor. Thus, the overload heating units used for running overload protection are rated at:

125% of 21 = 1.25 X 21 = 26.25 amperes

Section 430-22 (a) specifies that a branch circuit feeding an individual motor shall have conductors with a current capacity of not less than 125 percent of the full-load current rating of the motor. The wire size of the branch circuit for the 7 1/2-hp motor is determined as follows:

125% of 21 = 26.25 amperes

Therefore, a No. 10 AWG wire can be used, as verified in *Table 310-16*.

The size of the fuses for the feeder may now be found by adding:

40 + 15 + 15 = 70 amperes

This means that 70-ampere time-delay fuses may be used. *Section 240-20* requires that overcurrent devices be placed in each ungrounded conductor. If the utility company supplies three-phase current with one phase grounded, a three-wire (solid neutral), two-fuse disconnect switch must be installed. When all three phases are ungrounded, a three-pole, three-fuse disconnect switch must be installed.

If additional equipment is to be installed, 100-ampere time-delay fuses may be installed.

Three-phase Entrance Switch (*Sections 230-70* through *230-84*)

Section 230-70 of the Code requires each service entrance to be provided with a readily accessible means of disconnecting all conductors from the source of supply. This may be either a manually operated switch or a circuit breaker. In this installation, a service-entrance switch is used.

Section 230-78 requires that the service-entrance switch be externally operable so that the operator is not exposed to contact with live parts. Furthermore, this switch must plainly indicate both the open and closed positions, *Section 230-77*.

It can be seen that three 70-ampere time-delay cartridge fuses may be used with a 100-ampere, three-pole, three-phase service-entrance switch.

Section 230-72 (c) permits the installation of the service-entrance switch on either the supply or load side of the meter test cabinet. However, power companies in different localities have their own rules regarding the installation of the service switch. Some power companies require the service-entrance switch to be connected on the supply side of the meter. Other companies require the connection of the service-entrance switch on the load side of the meter.

Meter Test Cabinet

Figure 13-1, page 86, shows a typical connection arrangement for a meter test cabinet on a three-phase, three-wire, 230-volt service entrance.

A special meter test cabinet is used for three-phase, three-wire service entrances up to and including 100 amperes. This cabinet is furnished by the power company. It is then the electrician's responsibility to install the cabinet and connect the supply wires and load wires at the test block. After the installation is complete and approved by the electrical inspector, the power company installs the watthour meter and makes the necessary connections between the meter and the test block.

Although power companies vary in their requirements, there are some general specifications that all companies require electricians to meet in their installations. A meter board must be installed for each indoor installation. The meter board must be

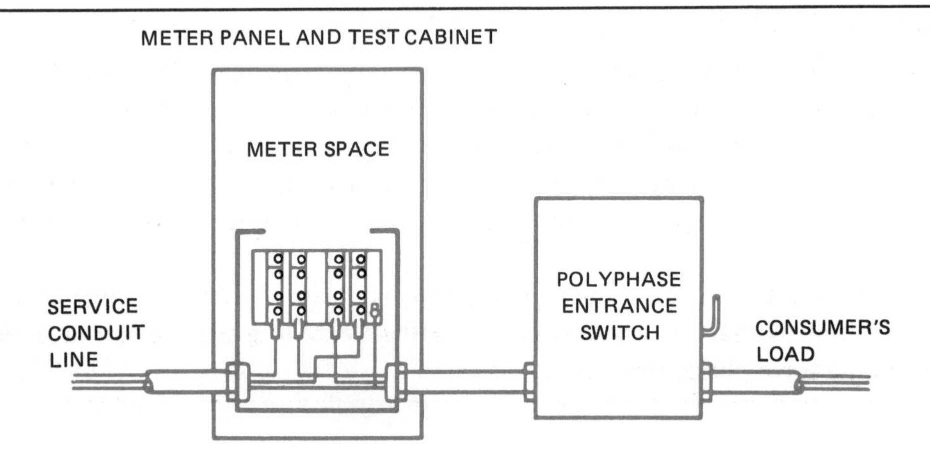

Fig. 13-1 Typical installation of a 100-ampere, 220-volt, three-phase, three-wire service entrance.

securely mounted in a true vertical position. For three-phase installations, there must be 24 inches of clear space for meter mounting above the meter test cabinet. If the meter board is located in a basement, the 24-inch clear space must be completely below the lower edge of any floor joists or other supports. At least six inches of clear space must be provided on all sides of the meter test cabinet. The top of the meter should be not less than four feet nor more than six feet from the floor.

Figure 13-2 illustrates the example given in this unit of a typical three-phase service installation. The service entrance is on the left and feeds into the meter entrance cabinet. Above the cabinet is the three-phase watthour meter which records the energy in kilowatt-hours. To the right of the meter entrance cabinet is the service-entrance switch. This three-pole switch is rated at 100 amperes, 230 volts, and is complete with 70-ampere cartridge fuses. From the service-entrance switch the feeder goes to the distribution box which contains the branch-circuit fuse protection. From the distribution box, the three branch circuits feed to the two 5-hp and one 7 1/2-hp motors in rigid conduit.

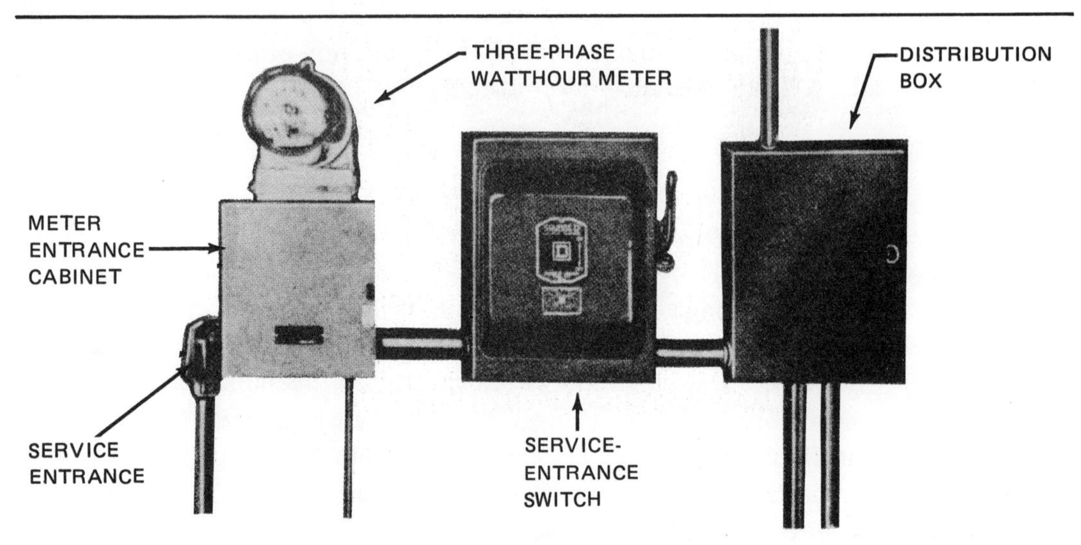

Fig. 13-2 A typical 100-ampere, 230-volt, three-phase, three-wire service entrance.

Three-phase Watthour Meter

The three-phase watthour meter, figure 13-3, is really two single-phase watthour meters mounted in the same case. Each single-phase unit has a separate voltage coil, current coil and disk. The disk turns or revolves between the current coil and the voltage coil. The disks of the two single-phase units are mounted on the same disk shaft and the total energy of the single-phase units is recorded in kilowatt-hours on the gear register.

The interconnection of these two single-phase units to record three-phase energy is exactly the same as that of two single-phase wattmeters being used to record three-phase power on a three-phase, three-wire system.

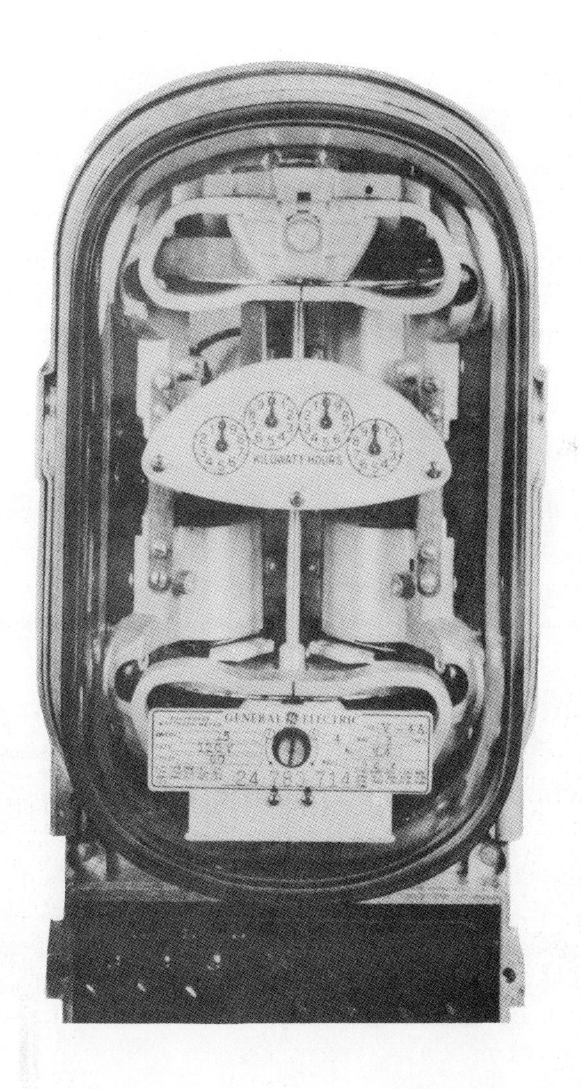

Fig. 13-3 A three-phase watthour meter. *(Courtesy General Electric Company)*

ACHIEVEMENT REVIEW

In items 1–5, select the *best* answer to complete the statement, and place the letter in the space provided.

1. One reason why three-phase service is used in preference to single-phase, for industrial motor loads is that __________

   a. installation is easier.
   b. three-phase motors are easier to maintain than single-phase motors.
   c. single-phase motors have better operating characteristics than three-phase motors.
   d. centrifugal switches are more efficient.

2. The watthour meter is installed by the __________

   a. power company.
   b. electrical inspector.
   c. fire marshal.
   d. plant foreman.

3. The three-phase watthour meter has built into it __________

   a. three single-phase watthour meters.
   b. a single wattmeter.
   c. two wattmeters.
   d. two single-phase watthour meters.

4. A watthour meter records __________

   a. power.
   b. watts.
   c. time.
   d. energy.

5. The service-entrance switch must __________

   a. indicate the open position only.
   b. indicate the closed position only.
   c. be externally operable.
   d. be operated internally.

6. Determine the wire size required for each of the three-phase motors listed below, and find the proper size feeder wire necessary to supply the motors as a group. Use TW wire.

   - One 230-volt, 27-ampere, 10-hp squirrel-cage motor with no Code markings
     Wire Size __________
   - One 230-volt, 15-ampere, 5-hp squirrel-cage motor with no Code markings
     Wire Size __________
   - One 230-volt, 9-ampere, 3-hp squirrel-cage motor with no Code markings
     Wire Size __________

     Feeder __________

7. Determine the maximum size time-delay fuses for branch-circuit protection for each motor in problem 6.

8. Determine the size of the fuse protection of the main feeder for the three motors in problem 6.

9. Determine the minimum size conduit required for each motor in problem 6, and find the size of the main feeder conduit.

10 hp	____________	in
5 hp	____________	in
3 hp	____________	in
Main Feeder	____________	in

14 Introduction to Fluorescent Lighting

OBJECTIVES

After studying this unit, the student will be able to

- explain the basic operating characteristics of fluorescent lighting units.
- list the basic circuit components for a simple one-tube fluorescent lighting unit of the preheat type.
- state what power factor conditions are caused by fluorescent lighting units.

ADVANTAGES OF FLUORESCENT LAMPS

Fluorescent lamps are used in great quantities for various installations. Because of its tubular form, the fluorescent lamp is called a tube. Fluorescent lamps range in length from 6 inches to 96 inches and have wattage ratings from 4 watts to 215 watts. Fluorescent lighting has several advantages.

- The wattage rating of a fluorescent unit of equivalent output in lumens is considerably less than that of the ordinary incandescent filament-type lamp. Two to four times as much light is produced per watt of power with a fluorescent unit as compared to the standard incandescent light bulb.
- Greater light output from a given circuit can be obtained without rewiring with larger size conductors if standard light bulbs are replaced with fluorescent units.
- The amount of heat given off by fluorescent lighting units is considerably less than that given off by incandescent light bulbs. This is an important factor to be considered when lighting air-conditioned buildings.
- The fluorescent unit has a low surface brightness. The fluorescent lamp or tube is not bright in one spot since the total light delivered by the tube is not dependent on a small area of extreme brightness but on a large area of relatively low brightness. As a result, there is better light distribution with fewer shadows and less eyestrain.
- Fluorescent lamps have a long life, and are available in various sizes, shapes, and colors.

FLUORESCENT TUBES OF THE PREHEAT TYPE

The *preheat* lamp is the earliest form of fluorescent lighting. This type of lamp is still in use to some extent. Therefore, the electrician will encounter this type of lamp and must understand how it functions, along with the related circuitry.

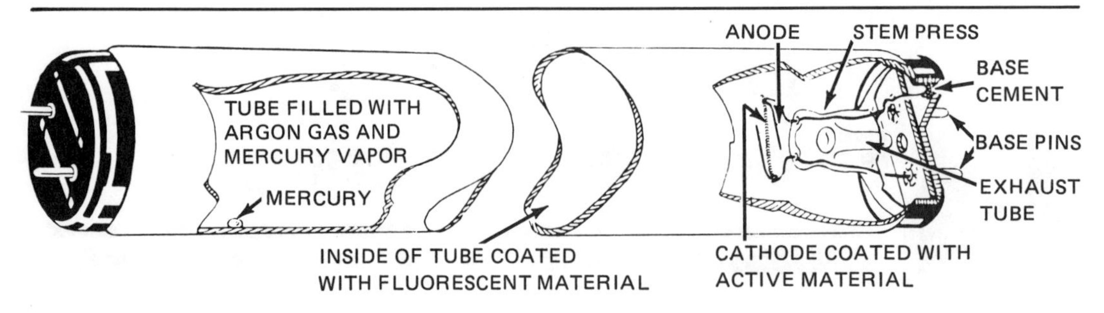

Fig. 14-1 Fluorescent tube. *(Courtesy General Electric Company)*

In the preheat type of lamp, each end of the fluorescent tube contains a cap with two terminals. Each set of two terminals or pins connects to a specially treated tungsten filament. The tube contains two filaments, one inside each end. The tube itself is filled with an inert gas and a small amount of mercury. Argon and argon-neon are commonly used gases, and krypton is used sometimes. The inside of the glass tube is coated with a chemical powder which will glow or fluoresce brightly when a current passes through the tube. By changing the mixture of the chemical powder, light of almost any color can be produced.

The details of the filament or cathode, the anode, and the terminals or pins are illustrated in figure 14-1.

Basic Circuit for a Preheat-type Fluorescent Tube

The connections for a fluorescent light unit consisting of one 15-watt, 115-volt, 18-inch tube are shown in figure 14-2. This tube requires special control equipment consisting of a ballast (series reactor) and a starting switch. The series reactor acts as a current-limiting device. The *starting switch* momentarily closes and opens the electrode heating circuit. This switch is also called a *glow tube, glow switch,* and *starter.*

When the circuit in figure 14-2 is energized, a small current passes through the series reactor, both tube filaments, and the glow tube. At the instant the circuit is energized, the current is very small because of the high resistance of the glow tube. The glow tube is a glass bulb or envelope filled with neon or argon gas. The tube also contains a U-shaped bimetallic strip, a 0.006-microfarad capacitor and a fixed contact or electrode. The capacitor eliminates any radio interference that may be caused by the opening and closing of the contacts. The contacts are open at the instant the circuit is energized.

Because of the high resistance of the glow tube, the current is small and there is little voltage drop across the series reactor. Therefore, there is sufficient voltage at the glow tube to produce a glow discharge between the U-shaped bimetallic strip and the fixed contact, figure 14-3. The heat from the glow causes the bimetallic strip to expand and close the contacts. Preheating takes place at both cathodes. The current through the two filaments is relatively high but the series reactor limits the current to a safe value. In the period that the contacts of the glow tube are closed, the temperature of the fluorescent tube electrodes rapidly increases. However, when the contacts close in the glow tube, the glow discharge is stopped, the bimetallic U-strip cools, and

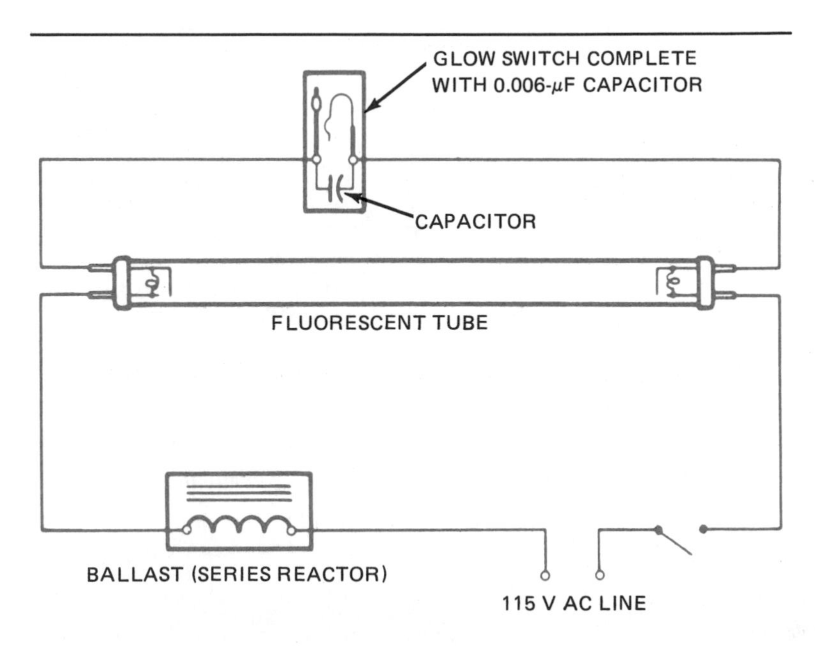

Fig. 14-2 Circuit for fluorescent tube.

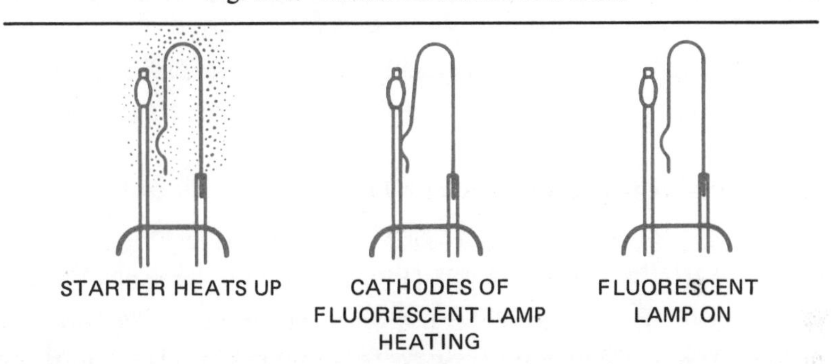

Fig. 14-3 A starting switch.

the contacts open. At the instant these contacts open, an inductive voltage kick generated by the series reactor coil starts conduction of current between the main electrodes of the fluorescent tube.

The fluorescent lamp continues to operate as long as the circuit is energized. The usual operating voltage for satisfactory operation is 110 to 125 volts ac. After the circuit is in operation, the reactor limits the current to the rated value so that the fluorescent tube fluoresces at the proper light intensity.

Power Factor Correction

The reactor or voltage ballast in series with the fluorescent tube causes the power factor of fluorescent units to range between 50 and 60 percent lag. The power companies, therefore, have requested the various fluorescent lamp manufacturers to install capacitors in fluorescent lighting fixtures to achieve power factor correction. Most fluorescent lighting units have such a capacitor connected in the lamp circuit so that the operating power factor of most fluorescent lamp units is near 100 percent or unity (1).

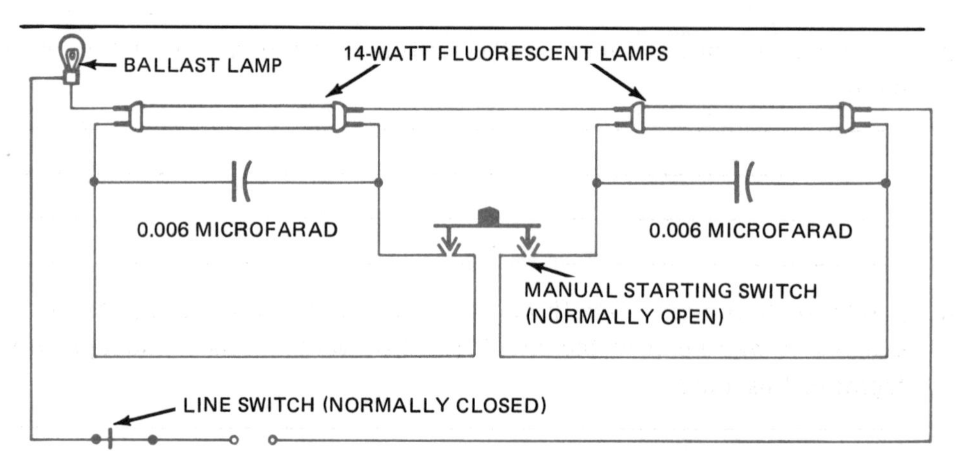

Fig. 14-4 A fluorescent tube circuit with manual starter.

The fluorescent tube circuit shown in figure 14-4 represents a circuit often used for desk-type fluorescent lamps.

A specially designed manual starting switch is used in this circuit. This switch has two functions:

1. When the ON pushbutton is depressed for a second or two and then released, the filament cathodes of the fluorescent tube are preheated.
2. Pressure on the ON pushbutton also closes contacts in the line wires. These contacts mechanically lock in the closed position. Pressure on the OFF pushbutton opens the contacts in the line wires to shut off the fluorescent lamp unit.

The two 14-watt fluorescent lamps shown in figure 14-4 are in series. A special small incandescent lamp is used instead of the traditional series reactor. This ballast lamp works very well as a current-limiting device. By using the ballast lamp instead of the typical series reactor coil, a fluorescent lamp unit has the following advantages: it is lower in cost, is lightweight, has no ballast noise, and has a high power factor value. The two small 0.006-microfarad capacitors eliminate any radio interference that may be caused by opening and closing the manual switch contacts.

ACHIEVEMENT REVIEW

1. List three advantages of fluorescent lighting units as compared to incandescent lamps.

   a. ______________________________

   ____

   b. ______________________________

   ____

   c. ______________________________

   ____

2. What gases are commonly used in fluorescent tubes?

   ____

3. What are two other names for the starting switch in a preheat fluorescent lamp circuit?

4. A 60-watt fluorescent tube fixture takes one ampere when connected to a 120-volt source. A wattmeter in the circuit reads 60 watts. What is the operating power factor of this unit?

5. A capacitor to improve the power factor is installed in the same 60-watt fluorescent unit given in question 4. The unit now takes 0.5 ampere at 120 volts. A wattmeter in the circuit reads 60 watts. What is the operating power factor of the unit?

6. When a capacitor is not used in a fluorescent lamp circuit, why is the power factor less than 1.0?

7. What is the purpose of the capacitor across the starter switch?

8. State two advantages of a ballast lamp as compared to a typical reactor coil.

   a.

   b.

15 Installation of Fluorescent Lighting

OBJECTIVES

After studying this unit, the student will be able to

- discuss the connections and operation of a few simple circuits used with preheat fluorescent lamps.
- tell how an instant-start fluorescent lamp operates and how it is connected in a circuit.
- explain the operation of a rapid-start fluorescent lamp.
- explain some of the maintenance problems and failures common to fluorescent lighting units.

The electrician is often required to locate and correct various problems in fluorescent lighting units. Faults can be quickly located and corrected if the basic principles of circuit operation are kept in mind. This unit covers some of the faults and failures common to fluorescent lighting units.

Part of this unit covers the circuits, control equipment, and operation of *preheat* fluorescent lamps. An explanation of the methods of correcting the power factor and eliminating the stroboscopic effect in these fluorescent fixtures is also covered.

The *instant-start* fluorescent lamp was developed to overcome the delayed lighting that occurs with preheat units. Due to the wide use of instant-start units, it is important that the electrician be familiar with this type of lighting. Therefore, information is given in this unit on the construction, circuit connections, and operation of instant-start fluorescent lighting units.

The *rapid-start* lamp is the most recent development in fluorescent lighting. This unit lights faster than the preheat type, but not as fast as the instant-start type. However, the ballast is more efficient and smaller, for a given amount of wattage, than the instant-start lamp. The rapid-start lamp has low-resistance cathodes which are heated with low power losses. Rapid-start lamps are the most popular type of fluorescent fixtures for new installations. An important feature of this type of lamp is that it can be used in dimming circuits.

PREHEAT FLUORESCENT LAMPS

Single Lamp with Autotransformer Ballast

When the rated wattage of preheat fluorescent lamps increases beyond 20 watts, there is a marked increase in the length of the fluorescent tube as shown in the following table:

Tube Wattage	Length of Tube
15 watts	18 inches
20 watts	24 inches
30 watts	36 inches
40 watts	48 inches
100 watts	60 inches

With 30-watt and larger fluorescent tubes, the typical lighting voltage value, between 110 and 125 volts, is too low to cause conduction. It is necessary to use some method of stepping up the voltage to a value high enough to start conduction. The method used is an autotransformer-type voltage ballast. An illustration of this type of ballast connected to a 30-watt preheat fluorescent tube is shown in figure 15-1.

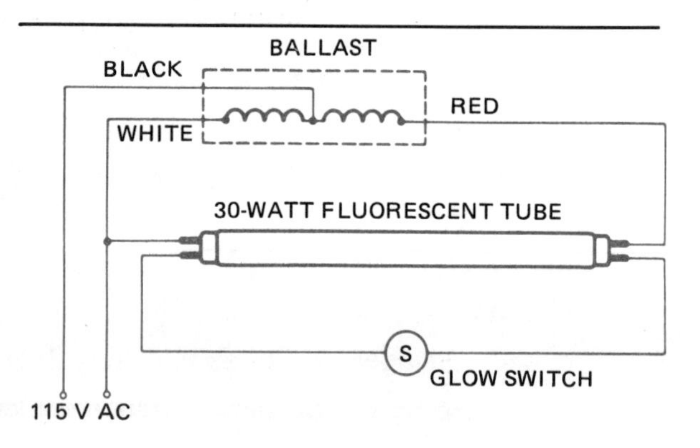

Fig. 15-1 Single lamp using autotransformer-type ballast.

The 115-volt input is applied to the section of the coil fed by the black and white line wires. The other section of the coil steps up the voltage output to the tube so that the tube starts conducting at the end of the preheat period. After the tube is operating, this section of the ballast acts as a series reactor to limit the current to the rated value.

Two-lamp Circuits for Preheat Fluorescent Tubes

The circuit connections for fluorescent lighting fixtures having two 15- or 20-watt tubes of the preheat type are illustrated in figure 15-2. The fluorescent tube marked tube No. 1 is in series with a reactor across the 115-volt source. The circuit for this tube is the same as the circuit for one tube covered in the previous unit. The starter switch marked "S" is a typical glow switch and is used to control the preheating of the lamp filaments of tube No. 1. The reactor coil for tube No. 1 is connected to the fluorescent tube by the blue conductor. The reactor coil for tube No. 1 has a large value of inductive reactance.

The second lamp, which is marked tube No. 2, is connected in series with a ballast consisting of a reactor coil and a capacitor. The connection between this second lamp and the ballast is made with a red wire. The path from the other side of tube No. 2 to the black line lead completes the circuit. The use of the capacitor in series with the reactor coil of tube No. 2 results in the current of this tube being as much as 90 to 120 electrical degrees out of phase with the current in tube No. 1. Because of this, the following is true:

1. the overall power factor is 95 percent or higher for the two lamps in parallel.
2. the stroboscopic effect is reduced considerably because of the phase displacement between the currents of the two lamps.

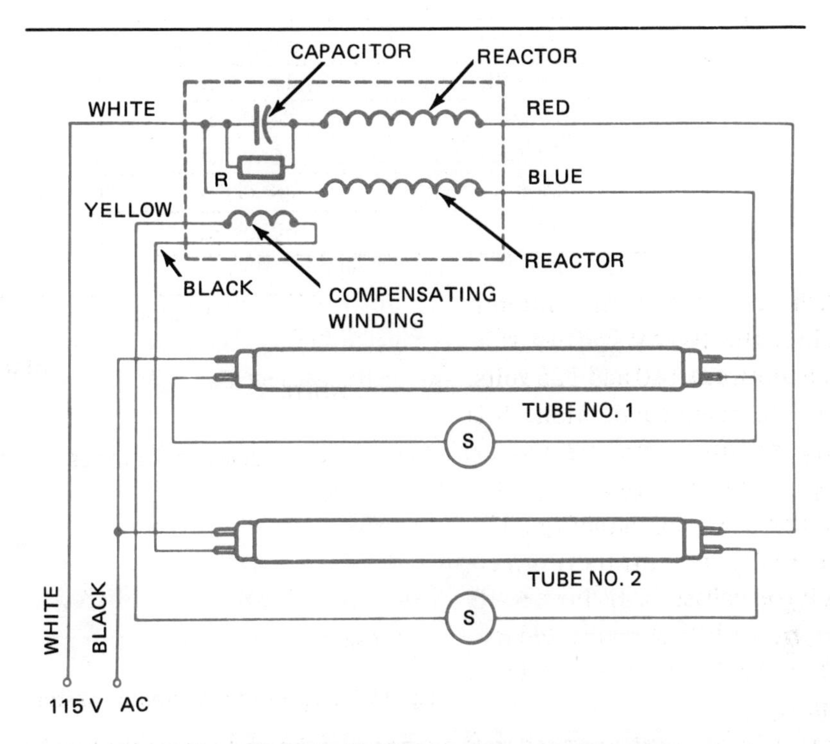

Fig. 15-2 Circuit for fluorescent fixture with two lamps.

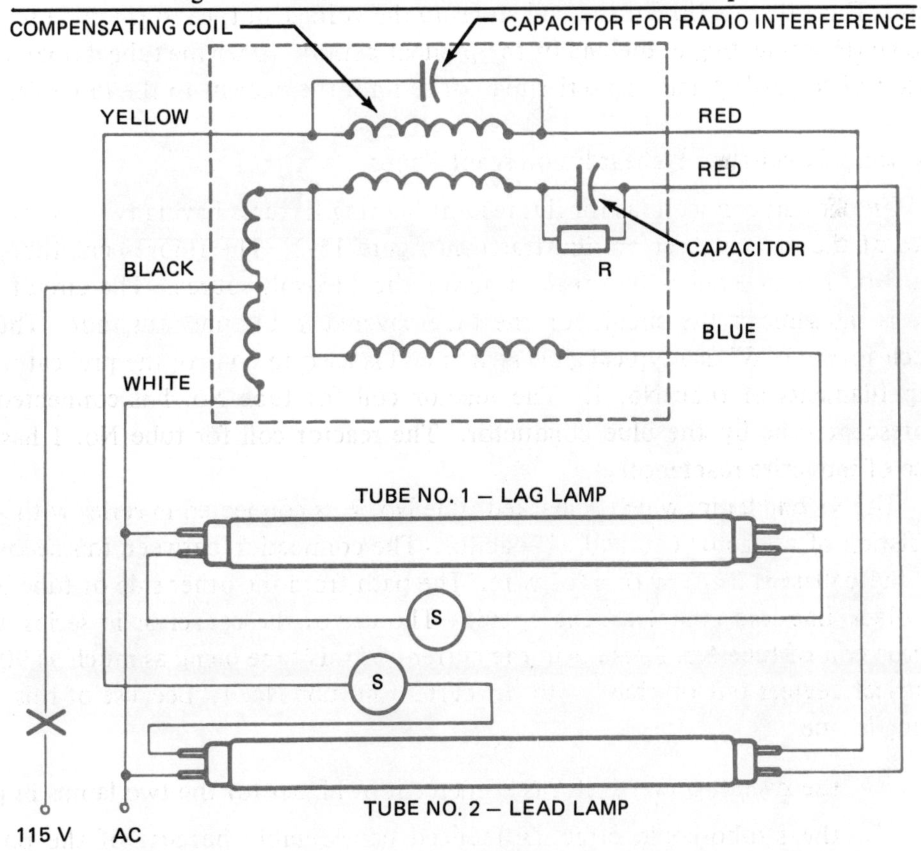

Fig. 15-3 Circuit for fluorescent fixture with two lamps of large capacity.

The capacitor in series with the reactor coil for tube No. 2 limits the inductive voltage kick of the coil. To start the tube, a compensating winding is used. The compensating coil is connected in series with the glow switch of this second tube by the black and yellow leads. The compensating coil gives the necessary additional inductive voltage kick to cause the second tube to start.

The circuit connections for fluorescent lighting fixtures having two 30-, 40-, or 100-watt preheat tubes are shown in figure 15-3, page 99.

The ballast used in figure 15-3 has an autotransformer to step up the line voltage. Wound on the same core with the autotransformer winding are the two reactor coils and a compensating coil. The autotransformer is necessary for the operation of 30-watt and larger fluorescent tubes.

Figure 15-4 illustrates the actual physical arrangement of the ballast for the circuit in figure 15-3.

The circuit using large-capacity lamps operates in the same manner as the two-lamp circuit for 15- and 20-watt fluorescent lamps, and it has the same desirable characteristics of high power factor and the reduction of stroboscopic effects.

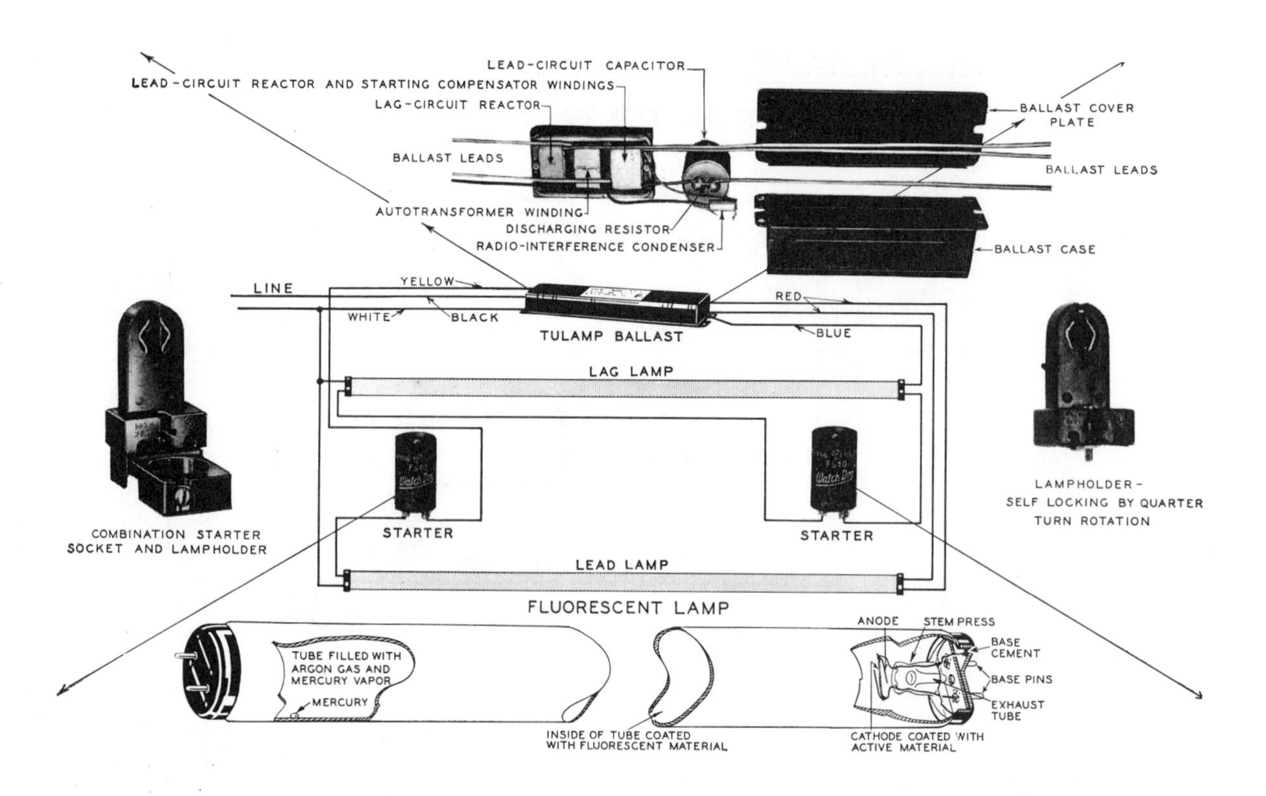

Fig. 15-4 Arrangement of component parts of a two-lamp circuit.

INSTANT-START SLIMLINE FLUORESCENT LAMPS

The Slimline fluorescent lamp, designed for instant starting, has only one terminal at each end. As in the preheat tube, the Slimline lamp has a filament type of cathode and operates as a hot-cathode lamp. The current passing between the two electrodes heats the segments of the small wire filaments to red-hot temperature in a fraction of a second. The Slimline lamp starts without preheating by using sufficient starting voltage. Therefore, the need for separate starters is eliminated.

Figure 15-5 illustrates the construction of the filament-type cathode and single terminal pin used on instant-start Slimline fluorescent lamps. This type of construction results in lower electrode losses. The other details of construction for instant-start fluorescent tubes are the same as for preheat fluorescent tubes, with the exception that the diameter of instant-start tubes is slightly smaller than that of preheat fluorescent tubes. (This characteristic promoted the use of the term Slimline.)

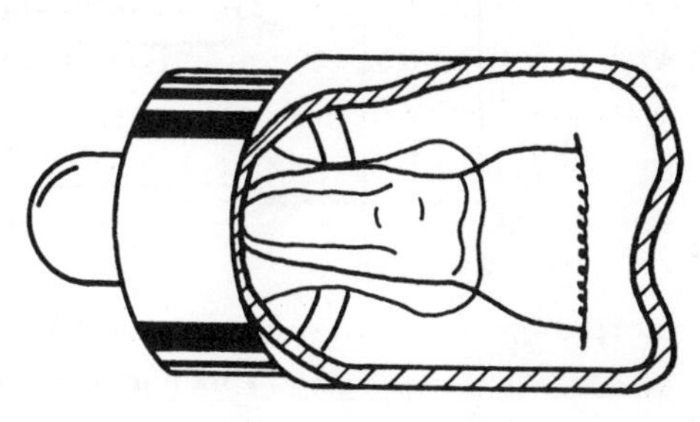

Fig. 15-5 A slimline single terminal pin.

Circuit for Instant-start Slimline Fluorescent Lamps

Figure 15-6 illustrates the connections for a circuit used to operate two instant-start Slimline fluorescent lamps.

The ballast used with this circuit is designed to:

1. deliver a high starting voltage at the instant the circuit is energized to start the lamps without preheating.
2. deliver a normal operating voltage after the lamps are in operation.

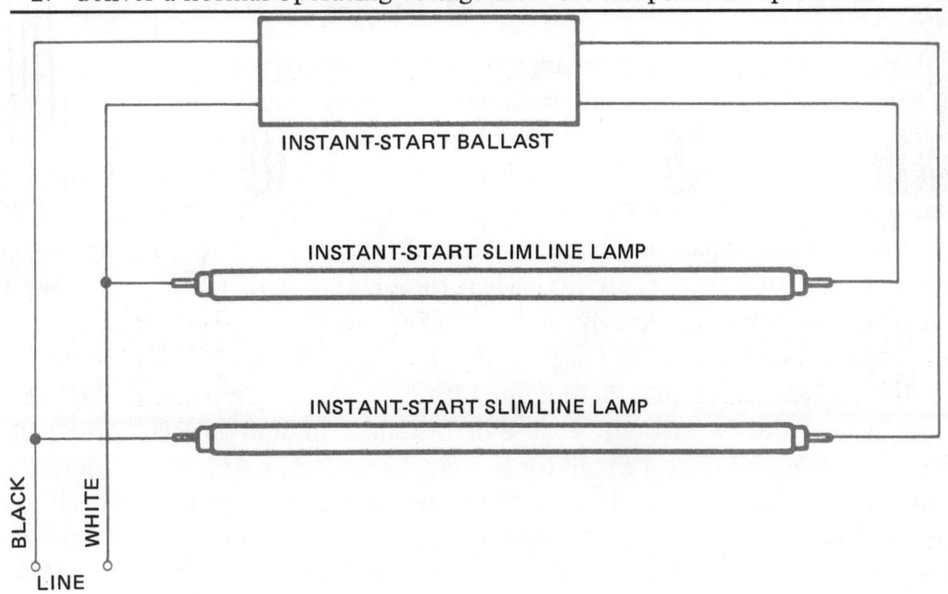

Fig. 15-6 Two slimline instant-start fluorescent lamps.

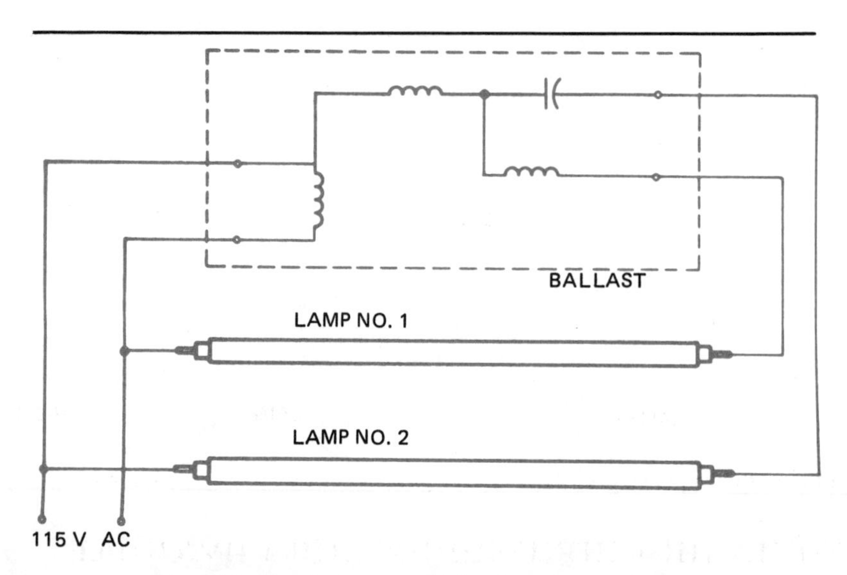

Fig. 15-7 Circuit for two slimline instant-start lamps.

When two-lamp Slimline circuits were first developed, they were of the lead-lag configuration with the lamps in parallel. Modern circuitry has the two lamps in series, figure 15-7, and the ballast is designed to start the lamps in very rapid sequence. This type of circuitry results in a smaller ballast, reduced cost, and lower sound level.

The use of instant-start circuits has the following benefits:

1. The resultant power factor for the two-lamp unit is 95 percent or higher.
2. The phase displacement between the currents in the two lamps reduces any stroboscopic effect.
3. This type of lighting unit starts the instant the circuit is energized.

RAPID-START FLUORESCENT LAMPS

The rapid-start lamp is widely used in modern installations. The cathodes are heated continuously, and the lamp is illuminated very quickly after the circuit is energized. The rapid-start lamp can be used in dimming and flashing circuits. Certain types of rapid-start lamps work very well in former preheat systems. A rapid-start lamp can be obtained for almost any type of weather condition. Figure 15-8 illustrates a fundamental rapid-start circuit. The ballast has separate windings to heat the cathodes continuously. Therefore, when the lighting switch is placed in the ON position, the lamps light very quickly, and no flicker occurs.

Since the cathodes are already heated, the amount of voltage required to cause the lamp to fluoresce is smaller than that required for the instant-start lamp. As a result, the rapid-start system is very efficient because of the small amount of loss in the ballast.

Figure 15-9 shows a typical circuit for two rapid-start lamps. It is a series circuit, and is commonly used. Once Lamp 1 is on, the voltage across it drops to a low value, and nearly all of the ballast voltage appears across Lamp 2. The starting voltage for this system is only a few volts higher than the voltage required to start one lamp. The result is that the size of the ballast can be small. In the rapid-start system, as in the instant-start system, there is no need for the separate starter and starter socket required in the preheat system.

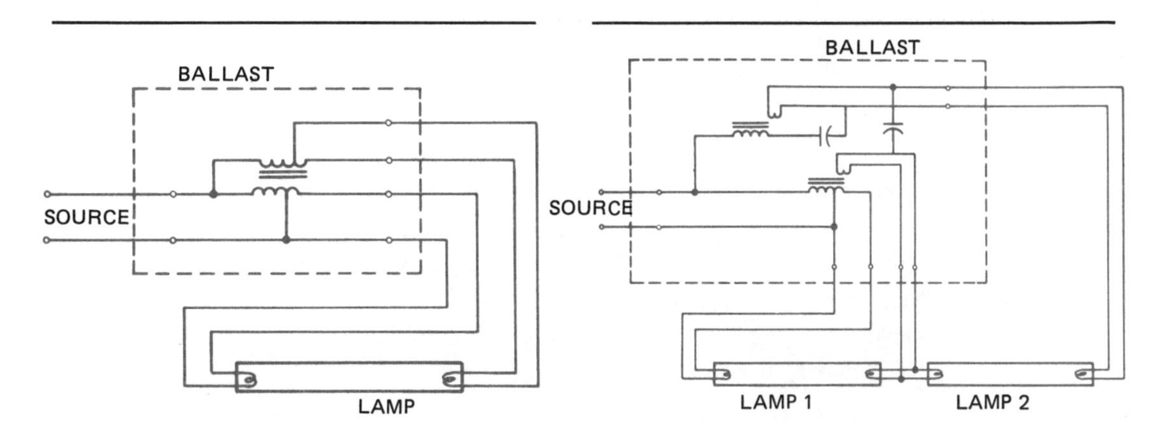

Fig. 15-8 Single-lamp, rapid-start circuit.

Fig. 15-9 Circuit for two rapid-start lamps in series.

GENERAL INFORMATION ON FLUORESCENT LAMP MAINTENANCE

To secure the best performance of fluorescent lamps it is important that the user understand how to properly maintain the installation. Certain factors affect the performance of fluorescent lamps which are not encountered with incandescent filament lamps. Some of these factors are within the control of the user.

To achieve maximum lamp life, it is suggested that lamps be operated continuously for periods of 3 to 4 hours. When a lamp is turned on and off frequently, the lamp life is decreased considerably. The filament cathode at each end of a fluorescent lamp is coated with a material from which there is an electrical emission when the unit is in operation. The material usually applied to the filament cathodes is a compound of barium or strontium. This material is gradually used up during the operation of the lamp and is consumed very rapidly during starting. The average life of most fluorescent lamps is approximately 7 500 hours, based on an operating period of three hours per start. Lamp life may be increased when the lamp is operated for longer periods of time.

Another point to consider is the temperature of the room in which fluorescent units are used. For satisfactory operation, the room temperature should be not less than 50 degrees Fahrenheit (50°F). If it is necessary to operate below 50°F, special fluorescent lamps can be obtained for use in low temperatures.

For preheat units, it is important that the correct starter be used. Unsatisfactory operation, consisting of extremely slow starting or flickering of the lamp unit, will result if the wrong type of starter is used. In some cases, the lamp unit may not even operate.

Fluorescent lamp units must be used with an alternating-current supply of the correct voltage and frequency. Check the specifications of the fluorescent lamp unit to insure that it is of the proper voltage and frequency rating. Also check the type of ballast used with the fluorescent lighting unit. Be sure that the ballast has the correct catalog or type number specified for the fluorescent lamp unit to be used.

Starting Difficulties

A fluorescent lamp may be difficult to start for a number of reasons. Any difficulty in starting results in a shorter lamp life. One common starting difficulty is that the lamp may blink on and off. This may mean that the lamp is at its normal point of failure. However, if the lamp is new or if it has been in service only a short time, a number of factors may be causing this condition. The difficulty may be due to the starter which, if this is the case, can be replaced readily. The lamp itself may be at fault. Check the lamp in another fluorescent fixture of the same size to determine whether the lamp is defective. Low circuit voltage, incorrect ballast rating, and low temperature also may cause the lamp to blink on and off. If a two-lamp fluorescent fixture is used, there is a possibility that the individual starter leads from the two pairs of lamp holders have been crossed. In this case, one lamp will start while the other lamp may blink on and off, or may not start at all. This sort of trouble can be located by simply removing one of the lamps from the lamp holders. With one lamp removed, the other will not start.

Another common starting problem is that the lamp makes no starting effort or starts very slowly. Be sure that the lamp makes proper contact in the lamp holders. The lamp can be checked by testing it in another circuit, as there may be an open circuit in the lamp filament. It is also possible that the starter has become defective and should be replaced. If necessary, a voltage check can be made with a voltmeter. If no voltage indication can be found, check the circuit connections including the circuit leads to the lamp holders. There is also a remote possibility that the voltage ballast may be open circuited.

Lamp Appearance

At the end of a normal life period, a fluorescent lamp usually shows a dense blackening at either or both ends of the tube. Little indication of blackening should occur during the first 500 to 1 000 hours of operation. When there is heavy end blackening during the first 1 000 hours of operation, the active material in the filament cathodes is being used too rapidly. This may be caused by:

1. frequent starting of fluorescent lamp units.
2. starters operating improperly, causing either too short or too long a preheat period.
3. improper ballasts which do not meet specification requirements.
4. circuit voltage being too low or too high. For best results, the circuit voltage should be within the rating of the fluorescent fixture unit.
5. improper wiring of the fluorescent fixture unit.

ACHIEVEMENT REVIEW

In items 1–10, select the *best* answer to complete the statement. Place the letter of the selected answer in the space provided.

1. The most recent development in fluorescent lighting is the ________
   a. preheat lamp.
   b. rapid-start lamp.
   c. instant-start lamp.
   d. glow tube.
   e. ballast lamp.

2. The most popular type of fluorescent lamp for new installations is the ______
   a. ballast lamp.
   b. instant-start lamp.
   c. preheat lamp.
   d. glow tube.
   e. rapid-start lamp.

3. The type of fluorescent lamp that can be used for dimming is the ______
   a. rapid-start.
   b. preheat.
   c. glow lamp.
   d. ballast lamp.
   e. instant-start.

4. The fastest starting and slowest starting lamps, respectively, are the ______
   a. rapid-start and instant-start.
   b. rapid-start and preheat.
   c. instant-start and preheat.
   d. instant-start and rapid-start.
   e. preheat and rapid-start.

5. The Slimline lamp is ______
   a. a rapid-start lamp.
   b. an instant-start lamp.
   c. a preheat lamp.
   d. a ballast lamp.
   e. a glow lamp.

6. A separate starter and socket is required for the fluorescent lamp type called ______
   a. instant-start.
   b. glow.
   c. Slimline.
   d. rapid-start.
   e. preheat.

7. The type of lamp that has the cathodes heated continuously before the lamp is to be lighted is the ______
   a. Slimline.
   b. rapid-start.
   c. instant-start.
   d. preheat.
   e. ballast.

8. The type of lamp that has a single terminal pin at each end is the ______
   a. rapid-start.
   b. preheat.
   c. glow.
   d. instant-start.
   e. filament.

9. Fluorescent lamps ______
   a. cannot be used in low-temperature environments.
   b. can use almost any type of starter.
   c. can be used with any type of ballast.
   d. should be operated continuously for periods of 3 to 4 hours.
   e. should be turned on and off as often as possible to increase lamp life.

10. If a fluorescent lamp shows a dense blackening at each end, even though the lamp was used correctly in a proper circuit with appropriate components, it means that ______
    a. the ballast is the wrong size.
    b. the voltage is too low.
    c. the lamp is at the end of its normal life.
    d. the fixture is improperly wired.
    e. the starter is the wrong size.

16 Summary Review of Units 11-15

OBJECTIVE

- To evaluate the knowledge and understanding acquired in the study of the previous five units.

In items 1–25, select the *best* answer to make each incomplete statement true. Place the letter of the selected answer in the space provided.

1. The minimum number of watts per square foot allowed by the Code in determining the lighting load for a single-family dwelling is __________
   a. 2. c. 5.
   b. 3. d. 7.

2. The voltage from one ungrounded conductor to another ungrounded conductor, in a normal three-wire installation for a single-family dwelling, is __________
   a. 115 volts. c. 230 volts.
   b. 150 volts. d. 460 volts.

3. The minimum number of appliance circuits permitted by the Code, for a single-family home, is __________
   a. 1. c. 3.
   b. 2. d. 5.

4. The minimum size AWG wire that is permitted for small-appliance circuits in a home is __________
   a. 8. c. 12.
   b. 10. d. 14.

5. For a single-family dwelling, the service ground wire is attached to the street side of the water meter __________
   a. for maximum safety.
   b. because it is an easy installation.
   c. only for certain types of homes.
   d. only when the service exceeds 50 amperes.

6. The minimum number of watts per square foot allowed by the Code in determining the lighting load for apartment dwellings is __________
   a. 1. c. 3.
   b. 2. d. 4.

7. The minimum load in watts that must be allowed for small appliances in apartment dwellings is __________
   a. 3 kW.
   b. 4 kW.
   c. 6 kW.
   d. 8 kW.

8. The minimum size feeder wire that can be used for a three-wire feeder supplying more than two two-wire branch circuits is __________
   a. No. 8 AWG.
   b. No. 10 AWG.
   c. No. 12 AWG.
   d. No. 14 AWG.

9. For a computed neutral load of 234 amperes, the minimum size of the RHW neutral wire is __________
   a. 250 MCM.
   b. 300 MCM.
   c. No. 1/0.
   d. No. 2/0.

10. The meter that records the amount of energy that is used is the __________
    a. wattmeter.
    b. watthour meter.
    c. voltmeter.
    d. ohmmeter.

11. An employee of the local power company is the only person who is permitted to install the __________
    a. bonding.
    b. branch circuits.
    c. power panel.
    d. watthour meter.

12. The minimum size TW wires required for a motor that has a current rating of 27 amperes is __________
    a. 1/0 AWG.
    b. No. 4 AWG.
    c. No. 8 AWG.
    d. No. 10 AWG.

13. The maximum size time-delay fuses for branch-circuit protection for the motor in problem 12 should be __________
    a. 40-ampere.
    b. 50-ampere.
    c. 100-ampere.
    d. 120-ampere.

14. The earliest type of fluorescent lamp is the __________
    a. rapid-start.
    b. instant-start.
    c. ballast.
    d. preheat.

15. The type of fluorescent lamp that can be used in flashing circuits is the __________
    a. rapid-start.
    b. instant-start.
    c. ballast.
    d. preheat.

16. Another name for the instant-start lamp is __________
    a. Thinline.
    b. ballast.
    c. Slimline.
    d. Thintube.

17. Fluorescent lamps can be used __________
    a. in temperatures above 50°F.
    b. only indoors.
    c. only in dry weather conditions.
    d. in almost any type of weather condition.

18. When fluorescent lamps are used to replace incandescent bulbs, usually ____________

    a. greater illumination occurs.
    b. less illumination occurs, but they look better.
    c. only one color can be obtained.
    d. more shadows occur, but there is less eyestrain.

19. When a fluorescent lamp shows a dense blackening at each end, it could possibly mean that the ____________

    a. wrong gas was inserted.
    b. lamp was started infrequently.
    c. starter is working improperly.
    d. lamp is new.

20. When a capacitor is used in a fluorescent lamp circuit to correct power factor, the power factor approaches ____________

    a. 0 percent.
    b. 0.5 percent.
    c. 80 percent.
    d. 100 percent.

21. The ballast lamp performs the same function as the ____________

    a. glow tube.
    b. reactor coil.
    c. starter.
    d. preheat fluorescent lamp.

22. One advantage of a preheat fluorescent lamp as compared to an incandescent lamp is that it ____________

    a. illuminates faster.
    b. provides less illumination per watt of power.
    c. generally has a longer life.
    d. can be used for heating as well as lighting.

23. Modern two-lamp, rapid-start circuits have the lamps connected ____________

    a. in series.
    b. in parallel.
    c. in a combination of series-parallel.
    d. to separate glow lamps.

24. A typical gas that is used in fluorescent lamps is ____________

    a. oxygen.
    b. nitrogen.
    c. xenon.
    d. argon-neon.

25. One of the reasons why the instant-start fluorescent lamp was developed was to ____________

    a. simply provide variety.
    b. overcome the delayed starting which occurs with the preheat type.
    c. provide the capabilities of dimming which could not be obtained with incandescent lamps.
    d. have a slower starting lamp than the rapid-start lamp.

26. A single-family dwelling has an active area of 1 700 square feet, and an 8-kW, 115/230-volt range is to be installed. Using the minimum number of watts per square foot permitted by the Code, determine the number of 115-volt, two-wire branch lighting circuits that are necessary.

27. The number of appliance circuits required for the dwelling in problem 26 is ______________________________.

28. Referring to problem 26, determine the size of TW wires for the oven branch circuit.

Acknowledgments

Sponsoring Editor
William W. Sprague

Senior Editor
Marjorie A. Bruce

Project Editor
Frances Larson

Reviewer
Dennis Gable

Title Page Photographs
Apprentice Committee
Joint Industry Board of the Electrical Industry
Flushing, New York

Index

681 (OC1745)